A STEP FARTHER OUT

'It is hard to over-praise this collection of articles on the latest developments in science, or their author. But I'm going to try. Of all the people I have ever met in my life, Jerry (Pournelle) is one of the most colossally educated in science. He had advanced training in systems engineering, psychology, physics, mathematics, logic, and political science. He is the only science degree person I have ever known how was able to explain coherently the entire Velikovsky controversy, an hour later do the same for the Atlantean legend, and then, in a major talk, describe the new, dynamic, holographic view of the structure of the brain.'
A. E. van Vogt

Jerry Pournelle worked on both the Mercury and Apollo space programmes, and was involved in the qualification tests for the original astronauts. He has degrees in engineering, psychology and political science, and is the winner of the 1972 John Campbell Award.

Also in *Star*:

A STEP FARTHER OUT

Part 1

by Jerry Pournelle, Ph.D.

Preface by Larry Niven
Foreword by A. E. van Vogt

Star

A STAR BOOK

published by

the Paperback Division of
W. H. ALLEN & Co. Ltd

A Star Book
Published in 1981
by the Paperback Division of
W. H. Allen & Co. Ltd
A Howard and Wyndham Company
44 Hill Street, London W1X 8LB

First published in Great Britain by
W. H. Allen & Co. Ltd, 1980

A Step Farther Out parts 1 and 2 were originally published as one volume.

Copyright © 1979 by Jerry Pournelle

Reproduced, printed and bound in Great Britain by
Hazell Watson & Viney Ltd, Aylesbury, Bucks

ISBN 0 352 30883 4

Dedication

For Jim Baen, an extraordinarily
good friend and editor.

Acknowledgments:

Research for this book was supported in part by grants from Pepperdine University and the Vaughn Foundation, to both of whom the author gives respectful thanks. Opinions in this work are the sole responsibility of the author.

It is obviously impossible to thank all those who have significantly contributed to a work this size. However, special thanks are due to Ejlar Jakobssen, who first encouraged me to do a regular science column; to Jim Baen, his successor as editor at *Galaxy*, who became invaluable for his topic suggestions and who unarguably improved the result; to Freeman Dyson, who generously granted ideas and interviews; to Dr. Gerald Yonas and Dr. John Penitz of Sandia Corporation; Dr. Petr Beckmann; Russell Seitz; A. E. van Vogt and Robert Bloch for their encouragement; Edmund Clay; and Larry Niven. Thanks are also due the members of the Los Angeles Science Fantasy Society, Inc., who were hardy enough to listen to my readings of these essays before they saw print.

It is also traditional for an author to thank his wife for putting up with him while he went through the emotional storms so often associated with writing. In my own case the thanks are for reasons considerably more valid than tradition.

JEP
Hollywood, 1978

Contents

Preface

The Freedom of Choice
by Larry Niven

Jerry Pournelle is out to make the whole world rich.

He's been at this for some time. Like a good many of his colleagues, Jerry was sucked into the space sciences by science fiction. He was building rockets for the government back when they had to steal the parts from other projects, and get the work done by sneaking back into the plant after clocking out. He's been building the future since I was in grade school, and he's still at it.

Of course, he would prefer to build it *his* way. Jerry has less of the ability to "suffer fools gladly" than anyone I know. (I'm not too good at that myself.)

His ambitions are impressive. In this book you'll find laid out for you several routes to a future in which the entire world is as wealthy as the United States is today . . . and that is as wealthy as any nation has been in human history. He does not intend that we should confine ourselves to Only One Earth.

Well, you'll get to that. Let me deal with another question. Do we *want* the whole world rich?

I happen to think we do, but I've heard other opinions.

Do you feel that your soul and body will benefit if you eat nothing but organically grown fruits and vegetables? You may well be right; but there's a reason why those scrawny carrots are so expensive. Without fertilizers and bug sprays the tomatoes, etc., might not come out of the ground. (Ours didn't!) Wealth lets you pay someone else to grow it. If you go the whole route, forming a commune, living as your ancestors did, eating only food you grow yourself without technological help ... then wealth lets you go on eating after the crop fails.

More generally, the right to live as if you were poor is inalienable. What you stand to lose is the right to live otherwise. Through your laziness or your inattention or through listening to the wrong saviors, you may condemn all future generations to involuntary poverty.

Nobody can be forced to spend wealth. That applies to you as thoroughly as it applies to the Indian rice farmer or Brahmin mendicant. Either can simply ignore the wealth that Dr. Pournelle proposes to drop on his head.

Granted that there are problems. A wealthy world would aggravate the servant problem no end.

Remember when people could sell themselves into slavery in order to eat? There was a ready market, because machines did not yet compete with musclepower. Those halcyon days are gone. With no good reason to fear for their jobs, servants have already become arrogant enough that most people would rather let a machine do it.

Well, why not? In the past few decades we've developed ultra-dependable ovens, vacuum cleaners, dishwashers, washer-dryers, soaps and detergents and other specialized chemicals for tasks each of which was once served by elbow grease (and somebody else's elbows, with any luck). The controls on my microwave oven have a better memory than my mother's cook, and my mother's cook quits more often.

In an age of inflation, the price of computer capability is going *down*. Ten years from now, your chauffeur may well be a computer; and why not? It would take up less room in the car and *far* less room in the house.

Consider backpacking. Over the decades, what was once a test of survival has become comfortable. Roads carry you into the wilderness. There you carry freeze-dried food and a lightweight mummy bag and air mattress on a contoured pack with a hip belt. Naturally the trails grow crowded. The population increases, the wilderness decreases. Already people propose to put glittering solar power collectors all over perfectly good deserts, instead of in orbit, as God intended.

If four billion people could afford to buy Kelty packs and sleeping bags, a certain minute percentage would go backpacking. And the world's wilderness areas would be jammed! What happens to the original backpacker, the man who needs the solitude of an empty trail?

No sweat. If we follow Jerry's route, we'll be moving a lot of our industries into Earth orbit and beyond. We'll be mining the Moon and the asteroids, and using free fall to keep heart patients alive and to manufacture ball bearings and single-crystal whiskers and strange new alloys. Let's continue that process. Move *all* of Earth's industries into Earth orbit. Turn the Earth into one gigantic park. There'll be room for the backpackers.

Does the world need to be rich? Suppose the worst: suppose none of the money is yours. What does the wealth of a society do for you?

The last time I spoke on this subject, someone in the audience called me a "bourgeois" for the first time in my life. Do we bourgeoisie tend to overemphasize wealth? Maybe. Someone else pointed out that, if we were all to spend most of our time in meditation, in seeking out the strengths and weaknesses of our own souls, we would use very little of the world's resources.

She was right, of course. I did have to point out that one would get the same benefits from being dead; but even that isn't the point. *Choice* is what matters. You have the right to choose your profession or lack thereof, your friends, when and whether you get married, what clothes you wear, how and whether to cut your hair and shave your face or legs,

11

and whether you spend twenty-four hours a day meditating. But that right depends absolutely on your ability to walk out! If the pressure from your parents or neighbors is too much, hop on a bus and go. Change cities, if necessary. You don't have to resist the pressure to conform. There are people living exactly as you would like to. Find them!

What does it take to maintain these freedoms? Not much. Classified ads in newspapers, a nationwide telephone network, your car and a network of highways and gas stations, several competing airlines, a public police force—actually a fairly recent invention, that one.

Fred Pohl's biography speaks of another freedom—a freedom you will hopefully never need. Fred grew up during the Depression, in a society that could not yet afford Welfare. There was no bottom to failure in those days. You could starve in the street, just like in India. Far and few were those willing to claim it was good for their own souls.

Oh, there's one more freedom worth considering, for those of the female persuasion. Laws tend to pragmatism. Your legal right to be considered the equal of a man depends on physical strength being irrelevant; and that depends on machines. Women have been slaves in most societies throughout most of human history. Sophisticated contraceptives help too; they allow you to avoid compulsory pregnancy. Peasants don't manufacture contraceptives.

If you're my age (forty) or younger, you've been living in a wealthy world for all of your life. Perhaps you haven't noticed. It's time. The sources of our wealth are running out. Dr. Pournelle will show you where to go for more.

Foreword
by A. E. van Vogt

It is hard to over-praise this collection of articles on the latest developments in science, or their author. But I'm going to try.

All of these pieces appeared originally in *Galaxy*, a major science fiction magazine. When the first one was printed in the April 1974, *Galaxy*, I was motivated to write the following letter to Jim Baen:

Dear Editor: I was pleasantly surprised when I finally opened your latest issue—the April—and saw Jerry Pournelle had done a science article for *Galaxy*. This is sf magazine science fact really moving up in the world. Of all the people I have met in my life, Jerry is one of the most colossally educated in science. He had advanced training in systems engineering, psychology, physics, mathematics, logic, and political science. He is the only science degree person I have ever known who was able to explain coherently the entire Velikovsky controversy, an hour later do the same for the Atlantean legend, and then, in a major talk, describe the new, dynamic, holographic view of the structure of the brain. Jerry has Isaac Asimov's memory in a younger body, and it comes out by instant association in a similar electrifying voice. Several sentences in his article referred to new developments in aerospace and energy science that were hitherto unknown to me—and I *try* to keep up. After reading that, the future already looks brighter to me. Scientifically oriented readers will now have to add *Galaxy* to their list of publications they need to keep up with what's going on.

That was published in the July 1974, issue. Now, here we are more than four years later. During the interval, Jerry has not only written a science article for every issue of *Galaxy*; he has also become a major visiting science lecturer at universities, and has gone up in the sf literary world to become—hear this!—one of the half dozen authors in the field who have received advances of more than a hundred thousand dollars. In 1973 he won the John W. Campbell, Jr. award as the most promising new sf writer of the year. In 1977, he and his collaborator, Larry Niven, were paid an advance of $236,500 for the paperback rights of their sf novel, LUCIFER'S HAMMER.

In these articles you will find that the author is pro-technology, pro-space program, pro-interstellar exploration. And he supports these and other pro-science projects for a strange reason. He can prove that they and they alone will accomplish what the anti-science proponents want. Without space colonies, the third world is dead. Without meteorite mining, people on welfare will presently get nothing.

Read Jerry's SURVIVAL WITH STYLE and A BLUEPRINT FOR SURVIVAL. We have (according to Jerry Pournelle) a hundred years to get out into space and save ourselves. After that, because of the depletion of the necessary resources down here on the surface of the planet—necessary, that is, for getting us off and up—the opportunity will never occur again for the human race.

A science article by Jerry Pournelle has an astonishing amount of writing energy in it. Like Isaac Asimov, Jerry puts himself out where you can see who's doing the reporting. Like Isaac he knows the facts and has the formal training to evaluate and present the data.

Another comment on that training: Jerry's mother once told a group of us that as the years went by, and there was Jerry still in college, taking this degree and that degree, and that training, and that one, and that, and that ... the family began to be worried.

As they, and we, may see, he came out of it all right. And not only as a brain. Jerry has a tall, lean, tough body which, in its time, served in Korea, achieved a high level of skill in the graceful, muscular art of fencing, and acquired the enduring heart and lung power that comes from hiking in the mountains.

Jerry has been an aide to a mayor of Los Angeles, a practising PhD psychologist, president (the same year winning that Campbell award) of the Science Fiction Writers of America, and other achievements.

As you have now seen, I've tried to over-praise Jerry Pournelle and what he has written in this book. But it can't be done.

Introduction

I want to show you marvels. Dreams, in technicolor, with sharp edges. I want to tell you something of the wonder and excitement of science; of the birth of the universe; of black holes, and cities of the future; of how man and computer may forge between them something greater than both; of the world of energy, from garbage to outer space; of worlds transformed, and how many may direct the evolution of stars. I want to show you a world that might be made.

I also warn you: you will be asked to make some decisions. They are not decisions you can avoid; you are making them now, every day of your lives; but you may not know just how important is your generation.

We live in an age of marvels. Despite that, we feel a sense of impending doom. When I was an undergraduate I was involved with dramatic arts, and at one time was assistant director of a theater. In due course I was given original plays to read with a view to producing them.

One was memorable only for its title: "First Document of the Last Generation."

"It seems to me, then, that by 2000 AD or possibly earlier, man's social structure will have utterly collapsed, and that in the chaos that will result as many as three billion people will die. Nor is there likely to be a chance of recovery thereafter."

Thus closes a popular article by Dr. Isaac Asimov, perhaps the best-known science writer in America, written even as Neil Armstrong set foot on the Moon. Not long after the Eagle landed, there was held in Stockholm a world conference entitled "Only One Earth"—and the consensus of the meeting was that "Spaceship Earth" was very likely doomed.

We are afraid of the future. At a time when the marvels of science are spreading throughout the world, when communications satellites bring hygiene and learning to the farthest corners of the globe, we have lost confidence in science and technology; indeed, in our abilities to control our world.

The intellectuals cry "Doom!" and many of us heed the warning.

Perhaps this is as it should be. Perhaps this loss of confidence in the ability of man to master his environment is justified. Perhaps we do not "have dominion" over the Earth, and the message of the doomsayers, of the advocates of "Zero-Growth," of those who put their faith in "small systems" and "soft energies" and "appropriate technologies" is no more than good sense.

Yet it does not seem obvious, at least not to me; and surely we ought not take such a momentous decision without thought of the consequences.

For make no mistake: the consequences of anything like Zero-Growth, of abandoning the idea of progress, are real and profound, and may not be reversible. The decision may be irrevocable, through all ages of ages.

And that is the reason for this book. I frankly reject the counsels of doom. I do not do so blindly. I am as aware as anyone of the dangers we face in the coming decades. I have certainly studied the problems, and it would be foolish to ignore the warning signs; mindlessly to proceed as we have done in the past would be disastrous—but I think it no less disastrous to cry "Doom!" and abandon faith in both technology and progress.

Of course this book does more than examine the alternatives we face. Over the years I have written of the fascination and excitement of science, and I have hoped to convey some of the sense of wonder—the conviction that we live in a generation of wonders—that I have felt. Perhaps that is the more important part of the book; yet I cannot help thinking that it is a worthwhile effort to show that we are not faced with doom; that the West need not close down; that we will survive.

I hope to convince you that this is the most important generation that ever lived; that what this generation decides will affect man's future every whit as much as did the development of agriculture, the invention of the wheel, the harnessing of fire; indeed, perhaps, as much as the evolution of lungs.

That is no small thing.

PART ONE:

SURVIVAL WITH STYLE

COMMENTARY

These essays were written over a period of two years, but they all deal with the same theme: can Western Civilization, and Mankind for that matter, survive? And if we do, will it be worth the effort?

The view that we are doomed has taken over a large part of the American intellectual community, and has been passed on to a generation of students. If accepted, it is a profound change in the traditional philosophy of the West.

According to FUTURE SHOCK, we are afraid of our future. It remains to ask—should we be? There is another view: that we can not only survive, but survive with style.

I am indebted to my editor, Jim Baen, for the title of this section and the lead column, as well as for his editorial assistance both during and after these were written.

Survival with Style

Suddenly we're all going to die. Look around you: a spate of works, such as THE DOOMSDAY BOOK, ECO-DOOM, and the like; and organizations such as "Friends of the Earth," and "Concerned Citizens" for one cause or another. All have the same message: Western civilization has been on an energy and resources spree, and it is time to call a halt.

The arguments are largely based on a book called THE LIMITS TO GROWTH. Written by a management expert for a group of industrialists calling themselves The Club of Rome, LIMITS TO GROWTH may be the most influential book of this century. Its conclusions are based on a complex computer model of the world-system. The variables in the model are population, food production, industrialization, pollution, and consumption of non-renewable resources. The results of the study are grim and unambiguous: unless we adopt a strategy of Zero-Growth and

22

adopt it now, we are doomed. Western Civilization must learn to make do, or do without; unlimited growth is a delusion that can only lead to disaster; indeed, *any* future growth is another step toward doom.

Doom takes any of several forms, each less attractive than the others. In each case population rises sharply, then falls even more sharply in a massive human die-off. "Quality of Life" falls hideously. Pollution rises exponentially. All this is shown in Figure 1, which is taken from one of the computer runs.

According to Meadows and many others, Earth is a closed system, and we cannot continue to rape her as we have in the past. If we do not learn restraint, we are finished.

Nor can technology save us. Perhaps the worst tendency of the modern era is our reliance on technologic "fixes," the insane delusion that what technology got us into, it can take us out of. No; according to the ecodisaster view technology not only will not save us, but will hasten our doom. We have no real alternative but Zero-Growth. As one ZG advocate recently said, "We continue to hold out infinite human expectations in a finite world of finite resources. We continue to act as if what Daniel Bell calls 'the revolution of rising expectations' can be met when we all know they cannot."

Jay Forrester, whose MIT computer model was the main inspiration for the zero-growth movement, goes much further. Birth control, he strongly implies, cannot alone do the job. It is a clear deduction from Forrester's model that

---

The "standard" model of World Three. The projection assumes no major changes in the physical, economic, or social relationships (as modeled in World Three). Population growth is finally halted "by a rise in death rate due to decreased food and medical services." "THE LIMITS TO GROWTH"

Figure 1

FOUR DOOMS
AS POSTULATED BY THE MIT
WORLD MODELS

FAMINE ··· POLLUTION ... OVERCROWDING ...

DEPLETION OF NON-RENEWABLES

Figure 2

. . . AND THIS IS THE WAY THE WORLD ENDS. . .

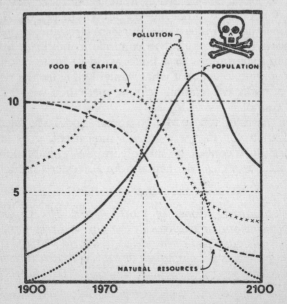

only drastic reductions in heatlh services, food supply, and industrialization can save the world-system from disaster.

It is important to recognize the severe consequences of a policy of Zero-Growth. For Western civilization ZG means increasing unemployment and a falling standard of living; worse than inconvenient, but not quite a total catastrophe. For the rest of the world things are not so simple. Behind all the numbers and computer programs there is a stark reality: millions in the developing countries shall remain in grinding poverty—forever.

They may be unwilling to accept this. There is then the decision to be made—must they be forced to accept? The advocates of Zero-Growth also advise, on both practical and moral grounds, the massive sharing of Western riches with the developing world. Indeed, under the ZG strategy, the West has only two choices: massive sharing with the developing world, or to retain wealth while most of the world remains at the end of the abyss. Neither alternative is attractive, but there is nothing for it: failure to adopt Zero-Growth is no more than selfishness, robbing our children and grandchildren for our own limited and temporary pleasures.

So say the computers.

*　*　*

I don't accept that. I want Western civilization to survive; not only survive, but survive with style.

I want to keep the good things of our high-energy technological civilization: penicillin, stereo, rapid travel, easy communications, varied diet, plastic models, aspirin, freedom from toothache, science fiction magazines, libraries, cheap paperback books, Selectric typewriters, pocket computers, fresh vegetables in mid-winter, lightweight backpacks and sleeping bags—the myriad products that make our lives so much more varied than our grandfathers'.

Moreover, I want to feel right about it. I do not call it survival with *style* if we must remain no more than an island of wealth in the midst of a vast sea of eternal poverty and misery. Style, to me, means that everyone on Earth shall have hope of access to most of the benefits of technology and industry—if not for themselves, then certainly for their children.

This is a tall order. Economists say it cannot be done. My wishes are admirable but irrelevant. The universe cares very little what we want; there are inherent limits, and the models of the world-system prove that what I want cannot be brought about.

That, however, is not so thoroughly proved as all that. Computers and computer models are very impressive, but a computer can give you no more information than you have put into it. It may be that Forrester and the other ecodoomsters have modeled the wrong system. At least it is worth taking a look; surely it is against man's very nature simply to roll over and die without a struggle.

Arthur Clarke once said that when a greybearded scientist tells you something is possible, believe him; but when he says it's impossible, he's very likely wrong. That has certainly been true in the past. Surely we are justified in examining the assumptions of those models which tell us we are doomed, and which dictate a policy of Zero-Growth.

* * *

The economists' models warn of four dooms: inadequate food supply; increasing pollution; depletion of non-renewable resources; and over-crowding through uncontrolled rise in population. Let us examine each in turn.

The first, food production, is surprisingly less critical than is generally supposed. This is hardly to deny that there is hunger and starvation in the world. However, given sufficient energy resources, food production is relatively simple. The UN's Food and Agricultural Organization reports that there are very few countries that do not, over a ten-year average period, raise enough food to give their populations more than enough to eat.

There are two catches to this. First, even in the West, birds, rodents, and fungi eat more of man's crops than ever does man. True we harvest more than most nations; but to do so requires high technology.

The second catch is the "over a ten year period" part. The *average* crop production is sufficient, but drought, flood, and other natural disasters can produce famine through crop failures over a one, two, or three year period. In much of the world there is no technology for storing surplusses. The West has known for a long time about the seven fat years followed by seven lean years, but it took us centuries to come up with reliable ways to meet the problem of famine.

Our solutions have been three-fold: increased production; better food storage, including protection from vermin; and weaving the entire West into a single area through efficient transportation. Drought-stricken farmers in Kansas can be fed wheat from Washington state, beef from the Argentine, and lettuce from California.

All this takes industrial technology on a large scale. Western farming methods use fertilizers. The transportation system is clearly a high-energy enterprise. Even providing mylar linings for traditional dung-smeared grain storage pits (animal dung is often the only waterproofing material avaiable) requires high-energy technology.

And in the West we waste land because we have land to waste; our agricultural technology produces surplusses.

A hard-working person needs about 7000 large Calories, or 7 million gram-calories, per day. The sun delivers nearly 2 gram-calories per square centimeter per minute; assume about 10% of that gets through the atmosphere, and that the sun shines about five hours (300 minutes) per day on the average. Further assume that our crops are about 1% efficient in converting sunlight to edible energy. Simple multiplication shows that a patch 35 meters on a side will feed a man—about a quarter of an acre.

Granted, that's an unfair calculation; but it isn't that far off from reality. My greenhouse, 2.5 meters on a side, can produce enormous quantities in hydroponics tanks, and there's no energy wasted in transportation and distribution of the food. I do use electricity to run the pumps, but that could be done, if necessary, by hand labor.

In Japan and in some of the oil-rich sheikdoms, hydroponics farming has been carried to fantastic lengths; acres of covered territory, with vegetables growing in the sandy deserts of Abu Dhabi, watered by desalinated seawater.

This is high-technology, of course. The chemical nutrients needed in my greenhouse take a lot of energy to manufacture. The greenhouse itself is made of aluminum tubing and mylar plastic reinforced with nylon strands. The piping and trays are plastic. All high-technology items, as are the fungicides I use, and even the water-testing kit that lets me balance off the pH in the nutrients.

Given the energy we can produce food. I think few would deny that. It is true enough that if the average Indian farmer could reach the productivity per acre achieved by the Japanese peasant of the 12th Century, India would have few food problems; but he's not likely to get there without industrial help (at the very least a television and satellite-relayed instructions). Moreover, the Japanese have had to move far ahead of their 12th Century output levels.

But I hope the point is obvious. Given sufficient energy, we have the technology to produce food. We may not have the energy; but famine is not a *primary* problem. With sufficient levels of industrialization we could even feed cities from greenhouses on the roofs of city buildings: if 1% of New York City were covered with greenhouses, they could feed 10% of the New York population. One percent of the surface area of Los Angeles would feed 1/3 that city's population.

We haven't even looked at the potential of the seas. True, our fish catches have about peaked out and may be declining—but man was never meant to be a hunter-gatherer.

Our exploitation of the seas is on a par with our use of land before we learned about agriculture and domestication of animals.

Sea-farming is a technology in its infancy; but experiments at St. Croix in the Virgin Islands (supported in part by the Vaughn Foundation which supported research for this book) show that fantastic levels of food production per acre can be achieved. The St. Croix research consisted of pumping cold nutrient-rich water from the sea bottom into pens where sunlight could energize plant growth; food harvested was shellfish and the like.

Other sea-farming enterprises in France and Britain show similar results. Selective fertilization of sea areas can increase sea-plant growth by orders of magnitude; one then introduces edible creatures which thrive on the plants. The production levels are again astounding, ten times what a given land area can produce.

Once again these are high-technology enterprises; but there is nothing far-out about them.

Clearly food production *per se* is not going to be a limit to growth for a very long time. Food production can only be limited by an enforced halt in industrialization and technology; given the energy, technology can easily feed far larger world populations than any projections anticipate for centuries.

* * *

If food production is not a primary problem, but rather an aspect of the energy shortage, pollution is doubly so. We already have the technology to clean up any and all pollutants.

It takes energy, of course. A lot of energy. But given the energy we can, if we must, take pollutants apart down to their constituent atoms.

The California Department of Public Health reports that the cleanest-running stream in the state is the outfalls of the Hyperion Sewage Disposal Plant for Los Angeles County. This is not a sad commentary on California's rivers; there are plenty of unspoiled streams in the High Sierra, but they do contain animal wastes from the deer and bears who inhabit the region.

I have on my desk a bottle of water taken from a sewage-treatment plant flowing into Lake Tahoe. Tahoe's problems are not technological; most of the water in the lake is reclaimed, and is indistinguishable from the cleanest mountain streams. True enough there are certain political jurisdictions which have not adequately cleaned up their act; but one must not blame technology for *that*. I use the Tahoe sewage water for ice cubes when I have a party that will have "concerned ecologists" as guests. It does no harm to show dramatically just how good our pollution-control technology can be.

Again I see no point in belaboring the obvious. *Given the energy resources,* pollution is not a real problem. Certainly pollution cannot be the limiting factor in industrial growth. It is another aspect of the energy shortage.

* * *

If famine and pollution do not define the limits to growth, then what of rising population? The view that we shall in the near future become so over-crowded that we will die of the resulting stresses is examined in detail in another chapter; for now let us look at the long-term prospects.

Throughout history there has been only one means of controlling population growth. It is not war; populations often rise in wartime. Famine and pestilence have of course reduced populations drastically, but the recovery from even these horsemen is often rapid, with birth rates skyrocketing so that within a generation population is higher than it was before the catastrophe. No: the only reliable means of limiting population is wealth.

The United States has a fertility rate below the replacement value; were it not for immigration the US population would begin to decline. (There is a "bow wave" effect from the WWII "baby boom" that distorts the picture, but the "boom babies" are rapidly reaching the end of their fertility epoch.)

France, Ireland, Japan, Britain, West Germany, Netherlands; where there is wealth there is decline in the birth rate. David Riesman in his THE LONELY CROWD pointed out many years ago that the Western nations were probably best described as in a condition of "incipient population decline," and it seems his prophecy was true.

Now it's true enough that if we manipulate exponential curves and thus mindlessly project population growth ahead, we will come to a point at which the entire mass of the solar system (indeed, of the universe) has been converted into human flesh. So what? It isn't going to happen, and no one seriously believes that it will. Obviously *something* will stop population growth *long* before that.

On a slightly more realistic scale, I have calculated how long it takes, at various growth rates, to reach "standing room only" on the Earth: that point at which there are four of us on each square meter of the Earth's surface (even counting the oceans and polar areas as "standable" surface), Figure 3 shows that those times are surprisingly near —if we have unlimited population growth. Yet the fact remains that as societies get wealthier, their ability to sustain larger populations increases—but their actual population growth declines or even halts.

Of course there are powerful religions whose adherents control large portions of the globe, and which condemn birth control and seemingly all other usable means of population limitation.

Yes. And I'm no theologian. But I cannot believe that any rational interpretation of scripture commands us to breed until we literally have no place to sit. Realistically we are *not* going to increase our numbers to that point: and, realistically, no religious leader is going to order it done.

"So God created man in his own image, in the image of God created he him; male and female created he them. And God blessed them, and God said unto them, Be fruit-

31

Figure 3

"So God created man in his own image, in the image of God created he him; male and female created he them. And God blessed them, and God said unto them, Be fruitful, and multiply, and replenish the earth, and subdue it; and have dominion over the fish of the sea, and over the fowl of the air, and over every living thing that moveth upon the earth."

Area, sphere: $A = 4 \pi R^2$
Radius, Earth: 6.371×10^8 cm.
Area, Earth: 1.700215×10^{18} cm^2

Standing room area requirement: 50 cm^2 = 2500 sq. cm.
  (About 4 people/sq. yard)

Number of people when Standing Room Only:
  6.80086×10^{14}
Present population: 4×10^9 (4 billion)

Assuming growth rate of 2% a year, it's SRO in 2584
At 1% growth, we get there in 3186 AD
At 4%, we get there in 2283.

  QUERY: At what point will the command be fulfilled?

ful, and multiply, and replenish the earth, and subdue it; and have dominion over the fish of the sea and the fowl of the air, and over every living thing that moveth upon the earth."

I will leave theology to the theologians; but the command was, "Multiply and replenish the earth, and subdue it;" and surely there must come a time when that has been *done?* When there can be no doubt that we have been sufficiently fruitful? And surely dominion over the wild things of the earth does not mean that we are to exterminate and replace them? Surely even those of the deepest faith may without blasphemy wonder if we are not rapidly approaching a time when we shall indeed have replenished and subdued the earth?

I cannot believe that we will continue to breed until we have destroyed our world; and frankly, I think of no more certain way to insure that the developing countries continue to increase in population than to condemn them to eternal poverty through Zero-Growth. So let's leave the bogeyman of unlimited population expansion. We have the technology to limit family size when, inevitably, there comes the time when everyone, no matter what his religous conviction, believes that the earth has been replenished and subdued.

Of course we have not reached that time: but the areas of uncontrolled population growth are the poorer areas of the world. All experience teaches that wealth will induce them to smaller family sizes, fewer children, control over population.

Wealth requires energy. The correlation between increase in Gross National Product and increased consumption of energy is about as well established as anything we know. There are those who search for exceptions—but they generally do not find them, and when they do there is always a very long "story" that goes with it. Common sense tells us that if we wish to become wealthy we will need the means of production; and productivity requires

Figure 4

LITTLE BUGS TO BIG BANG: SOME ENERGY EVENTS

Exponential Notation: 10^2 = 100, i.e., 1 followed by 2 zeroes. 10^3 = 1,000, 10^6 = 1,000,000, etc.

EVENT:	ERGS:
Mosquito taking flight	1
Man climbing one stair	10^9
Man doing one day's work	2.5×10^{14}
One ton of TNT exploding	4.2×10^{16}
US *per capita* energy use, 1957	2.4×10^{18}
Converting one gram hydrogen to helium	6.4×10^{18}
Saturn 5 rocket	10^{22}
One megaton, as in bombs	4.2×10^{22}
Total annual energy use, Roman Empire	10^{24}
Krakatoa	10^{25}
Annual output, total US installed electric power system, 1969	5.4×10^{25}
Thera explosion (largest single energy event in human history)	10^{26}
Total electric power produced, world, 1969	1.6×10^{26}
Total present annual energy use, world	10^{29}
One Solar Flare	10^{31}
Annual Solar Output	2×10^{39}
Nova	10^{44}
Quasar, lifetime output	10^{61}
BIG BANG	10^{80}

machinery, and that requires energy. Indeed, you could make the very definition of wealth the ability to dispose of great quantities of energy.

* * *

Thus we see that of our four dooms, three are aspects of the energy crisis: given sufficient energy we will not be overwhelmed by problems of food, pollution, or even over-population. But can we find the energy? Will not generating energy itself pollute the earth beyond the survival level?

At this point I must introduce some elementary mathematics. I will try to keep them simple and work it so that you don't have to follow them to understand the conclusions, but if I am to halfway prove what I assert I simply must resort to quantitative thinking. Failure to caluculate actual values, blind qualitative assertion without quantity, has been the genesis of a very great deal of misunderstanding and I don't care to add to that storehouse of misinformation. Besides, only through numbers can you get any kind of "feel" for the energy problem.

The basic energy measurement is the erg. It is an incredibly tiny unit: about the amount of energy a mosquito uses when she jumps off the bridge of your nose. In order to deal with meaningful quantities of energy we will have to resort to powers-of-ten notation. Example: $10^2 = 100$; $2 \times 10^2 = 200$; $10^3 = 1000$; and 10^{28} is 1 followed by 28 zeros.

Some basic energy events are shown in Figure 4. Note that a number of natural events are rather large compared to man's best efforts.

It takes a billion ergs to climb a stair, and a day's hard work uses 100,000 times more; yet a ton of TNT exploding contains a hundred days' work and more, while converting one gram of hydrogen to helium will yield more energy than each of us used in a year—and by "used" I don't mean each of us directly, but our share of all the energy used that year in the US: dams, factories, mines, automobiles, etc. I need hardly point out that there are a lot of grams (a gram is one cubic centimeter) of water in the oceans.

Nor need we worry about "lowering the oceans" when we extract hydrogen for fusion power. True, some rather silly stories have asserted that we might, but a moment's calculation will show that if we powered the Earth with each of 20 billion people consuming more energy than we in the US do now, the oceans would not be lowered an inch for some millions of years.

Of course fusion might not work. Given the present funding levels we may never achieve it, or the concept itself may be flawed, or the pollution associated with successful fusion may be unacceptable. Are there other methods?

One possible system is pictured in Figure 5. It is an Earth-based solar power system, and the concept is simple enough. All over the Earth the sun shines onto the seas, warming them. In many places—particularly in the Tropics—the warm water lies above very cold depths. The temperature difference is in the order of 50° F, which corresponds to the rather respectable water-pressure of 90 feet. Most hydro-electric systems do not have a 90 foot pressure head.

The system works simply enough. A working fluid—such as ammonia—which boils at a low temperature is heated and boiled by the warm water on the surface. The vapor goes through a turbine; on the low side the working fluid is cooled by water drawn up from the bottom. The system is a conventional one; there are engineering problems with corrosion and the like, but no breakthroughs are needed, only some developmental work.

The pollutants associated with the Ocean Thermal System (OTS) are interesting: the most significant is fish. The deep oceans are deserts, because all the nutrients fall to the bottom where there is no sunlight; while at the top there's plenty of sun but no phosphorus and other vital elements. Thus most ocean life grows in shallow water or in areas of upwelling, where the cold nutrient-rich bottom water comes to the top.

More than half the fish caught in the world are caught in regions of natural upwelling, such as off the coasts of Ecuador and Peru.

The OTS system produces artificial upwelling; the result will be increased plankton blooms, more plant growth, and correspondingly large increases in fish available for man's dinner table. The other major pollutant is fresh water, which is unlikely to harm anything and may be useful.

Certainly there are some engineering problems; but not so much as you might expect. The volumes of water pumped are comparable to those falling through the turbines at a large dam, or passing through the cooling system of a comparable coal-fired power plant. The energy itself can be sent ashore by pipeline after electrolysis of water into hydrogen and oxygen; or a high-voltage DC power line can be employed; or even used to manufacture liquid hydrogen for transport in ships as we now transport liquid natural gas.

As to the quantity of power available: if you imagine the continental United States being raised 90 feet, forming a sheer cliff from Maine to Washington to California to Florida and back to Maine; then pour Niagra Falls over every foot of that, all around the perimeter forever; you have a mental picture of the energy available in one Tropic, one band between the equator and the Tropic of, say, Cancer. It is more than enough power to run the world for thousands of years.

Finally the feasibility of OTS: in 1928 Georges Claude, inventor of the neon light, built a 20 kW OTS system for use in the Caribbean. It worked for two years. One suspects that what could be done with 1928 technology can be done in 1988.

OTS is not the only non-polluting system which could power the world forever. Solar Power Satellites would do the task nicely. SPS will be discussed in later chapters; but few doubt that they could provide more than enough

Figure 5

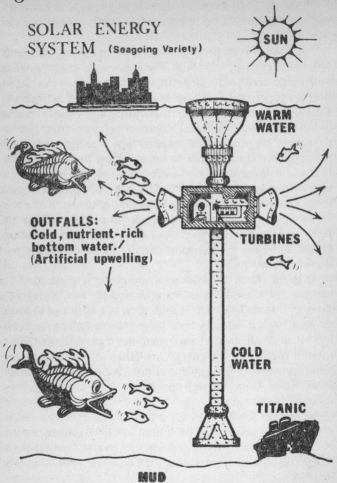

SOLAR ENERGY SYSTEM (Seagoing Variety)

SUN

WARM WATER

OUTFALLS:
Cold, nutrient-rich
bottom water.
(Artificial upwelling)

TURBINES

COLD WATER

TITANIC

MUD

energy to industrialize the world, and we understand how to build them far better at this moment than we understood rockets on the day President Kennedy committed us to going to the Moon in a decade.

That is a point worth repeating: we can power the Earth from space. We do not "know how to do it" in the sense that all problems are solved; but we do know what we must study in order to build large space systems. When John F. Kennedy announced that the United States would land a man on the Moon before 1970, the reaction of many aerospace engineers was dismay: not that anyone doubted we could get to the Moon, but those closest to the problem were acutely aware of just how many details were involved, and how little we had done toward building actual Moon ships. We had at that time yet to rendezvous or dock in space; there were no data on the long-term effects of space on humans; we had not successfully tested hydrogen-oxygen rockets; there were guidance problems; etc, etc. Thus the dismay: there was just so much to do, and ten years seemed inadequate time in which to do it.

Solar Power Satellites, on the other hand, have been studied in some detail; and we have the experience of Apollo and Skylab. We know that large structures can be built in space; they require only rendezvous and docking capabilities, and we've tested all that. We know we can beam the power down from space; the system has been tested at JPL's Goldstone, and the DC to DC efficiency was 85%. There are other problem areas, but in each case we know far more now than we knew of Mooncraft in 1961.

Ocean Thermal and Solar Power Satellites: either would power the world. I could show other systems, some not so exotic. My engineering friends tell me that OTS and SPS may even be the hard way, and there are much more conventional ways to supply Earth with energy.

No matter. My point is that *we can find the energy*. The method used is unimportant to the argument I make here: that we can survive, and survive with style.

Given energy we will not starve; we will lick the pollution problem; and we will generate the wealth which historically has brought about population limits. At least three of the dooms facing us can be avoided.

* * *

That brings us to the fourth doom: depletion of non-renewable resources. Can we manufacture the materials needed for survival with style? And can we do it without polluting the earth?

Surely we can. We can go to space to get the materials—and in doing it we can avoid pollution entirely. (There are, of course, those who worry about "polluting outer space", an example of non-quantitative thinking. Were we to devote the Gross World Product exclusively to the task and vaporize the Earth in the attempt we could not manage to pollute more than a fraction of a percent of the space in the solar system, and our effect would be temporary. One suspects that those who worry about "polluting outer space" are either incredibly arrogant, or actually are motivated by a desire for Zero Growth for its own sake.)

Metal production makes an excellent example. Mining and refining metals are some of the most polluting actions we manage, and metals are the most irreplaceable non-renewable resources we have. Give us enough iron and steel, copper, aluminum, zinc, and lead, and surely we'll have our problems licked. Give us enough metals and energy and we'll have wealth.

After all, it's mine tailings that produce some of the really horrible pollution; copper refineries that poison so many streams; and those belching steel mills that made Pittsburgh a legend (although Pittsburgh is also an excellent example of how pollution may be cleaned up once it is determined that cleanup has to be accomplished; a whole generation has never seen the smoke and fire of old Pittsburgh). Furthermore, processing metals uses up vast amounts of energy.

Give us metals free and clear, and the rest is easy. Give

Figure 6

METALS FOR THE WORLD....

In 1967, the United States produced 315 million tons of iron, steel, rolled iron, aluminum, copper, zinc, and lead.

Total metal produced, USA, 1967: 2.866×10^{14} grams.

Assume 3% ore, of density 3.5 gm/cm³,
    and the USA produced the equivalent of a sphere
    1.7 kilometers in diameter.

At 230,000,000 population, we produced 1.25×10^6 grams *per capita*. To supply the world with that much requires 5×10^{15} grams or FIVE BILLION TONS.

Assuming 3% ore at 3.5 gm/cm³, five billion tons of ore is a sphere 2.25 kilometers in radius
    or 4½ kilometers in diameter.

There are 40,000 or more asteroids larger than 5 km in diameter.

We may not run out of metals after all....

us enough metals and we'll industrialize the world. Besides, if we can do *that* in space, we can probably do anything else that has to be done. Consequently, I'll use metal production as my illustrative example.

In 1967, a year for which I happen to have figures, the United States produced 315 million tons of iron, steel, rolled iron, aluminum, copper, zinc, and lead. (I added up all the numbers in the almanac to get that figure.) It comes to 2.866×10^{14} grams of metal. Assume we must work with 3%-rich ore, and we have 9.6×10^{15} grams of ore, or 10.5 billion tons.

It sure sounds like a lot. To get some feel for the magnitude, let's put it all together into one big pile. Assuming our ore is of normal density we end up with a block less than 1.5 kilometers on a side: something more than a cubic kilometer, something less than a cubic mile. Or, if you like a spherical rock, it's less than two kilometers in diameter.

There are 40,000 or more asteroids larger than 5 km in diameter.

We may not run out of metals after all. . . .

But—the title of this chapter is "Survival with Style." Style to me does not consist of the West as an island of poverty in the midst of a vast sea of misery. Style, to me, means that everyone on earth has a chance at wealth—at least at a decent life.

Can we not agree that if everyone on Earth had the *per capita* metal production of the US, we would probably have achieved world riches? Especially since we export much of ours to begin with; surely it's enough?

Thus we take our 315 million tons and multiply by the world population, then divide by the US population; assume 3% ore, and we find how much we'll need. The result works out to a sphere less than four miles in diameter —and there are well over 100,000 asteroids larger than that.

Three percent ore is no bad guess as to what they're

Figure 7

CALL SMYTHE, THE SMOOTHER MOVER...

Take one each, FIVE BILLION TON asteroid.
Move from the Belt to Earth orbit.

Requires a velocity change of 7 kilometers a second.

$$KE = \tfrac{1}{2} M V^2$$

or, we need 1.225×10^{27} ergs.

For reference, the world annual energy use is 10^{29} so we're using about 1% of it. . . .

That's also 30,000 megatons.
And 30,000 one megaton bombs might just do it.

For a slightly more efficient system,
we can get the energy by converting 2,000 tons of hydrogen to helium...

Once we have the rock in Earth orbit, it's simple to get the metal out. We merely boil the entire rock.
Of course that takes rather large mirrors, but what the heck. . . .

---

made of, either. Actually, given the data from the Moon rocks, 3% is an underestimate of the usable metal content of the average asteroid. We've had heavy nickel-iron meteorites fall that were nearly 80% useful metal. Then too, some of the asteroids were once differentiated—that is, they were large enough that metallic cores formed. Then over the last four billion years the planetoids got bashed around until a lot of the useless exterior rock was knocked away, leaving the metal-rich cores exposed where we can get at them.

Over 100,000 asteroids, each capable of supplying the world with more metal per person than the US consumes in a year. Surely we won't run out of metals—but can we use them?

Sure we can. First, for the moment let's forget that the asteroids are *way* out there in the Belt, and concentrate on how to get the metals out assuming we have the rocks in Earth orbit. That turns out to be easy. We can use sophisticated methods, but there's also brute force: boil the rock.

It takes about 2000 calories per gram to boil iron. That's about the worst case for us, so we'll imagine the entire asteroid is made of iron. It takes, then, about 8.8 x 10 ergs, or twenty thousand megatons, to boil it all away.

The sun delivers at Earth orbit about 1.37 million ergs a second per square centimeter, and out in space we can catch that with mirrors. To boil our rock we could put up a mirror 80 kilometers in radius. That's too big; but we don't have to boil it all at once. A much smaller mirror to focus the sun onto a small part of the rock would be preferable.

A space mirror need be nothing more than the thinnest aluminized mylar, spun up to keep its shape. There's no wind or gravity in space. A mirror one or two kilometers across is a relatively simple structure—and more than adequate for our job. If need be we can actually *distill* off the metals we want.

Note, by the way, that there's been absolutely no pollution of Earth so far—even though we've got metals for the entire world. All the waste is out in space where it can't hurt us. But we do have a problem. My metals are *not* in Earth orbit; they're out there in the asteroid Belt, and they've got to be moved here—and that's going to take *energy.*

Let's see just how much it does take. To get from Ceres to Earth you need a velocity change of about 7 kilometers a second. By definition energy is mass given a velocity change, so we can quickly figure out how much; if we move the entire rock it comes to about 1% of the world's energy budget. That's not so much; we expend far more than that on metal production already.

To be more precise, it's about 60,000 megatons; and if need be, we can use hydrogen bombs. Put an H-bomb at the center of mass of an asteroid and light it off; I guarantee you that sucker will *move*. It's expensive, but not grossly so, assuming I have laser triggers for my H-bombs; only a few tons of hydrogen.

I could also do it with fusion: at 10% efficiency I get 6.4 x 10^{17} ergs per gram of hydrogen, and I need about 10^{27} ergs total to move the rock; for an engine I use an ion engine, breaking up parts of the asteroid for reaction mass. What arrives is something less than I started with, but who cares? What I'll throw away as reaction mass is the slag from my refinery.

(For those who haven't the foggiest notion of what I'm talking about: a rocket works by throwing something overboard. The reaction mass is what's thrown. Although the big space program rockets use gaseous exhaust as reaction mass, there's no reason you couldn't use dust, ground up rock, or slag from a metals refinery. It's all a question of whether you can throw it sternwards fast.)

But that leads to another possibility: why not set up the refinery out at the Belt? Put up vast mirror systems and do the refining on the way in; use the slag as reaction mass to move the whole works, rock, refinery, and all. I can power *that* with solar mirrors. Or I can do all at once: use bombs for initial impetus, set up mirrors when I'm closer in, and while I'm at it run a hydrogen fusion plant aboard the moving strip-mine/refinery/spaceship I have created.

At worst I have to carry about one Saturn rocket's worth of hydrogen, plus several shiploads of crew and other gear;

and for that I get an entire *year's* worth of metals for the world. The value of my rock is somewhere near a trillion dollars once it's in Earth orbit; more than enough to pay for the space program and pay off the National Debt at the same time.

So. For the price of some hydrogen and a rather complex ship system I've brough home enough metal to give everyone on Earth access to riches. If we do nothing else in space; if we come up with no new and startling processes such as I've described in other columns (and in other chapters of this book)—we'll have licked pollution and dwindling resources, thereby letting the developing countries industrialize, and thereby whipping the food production crisis for a while.

We have avoided the fourth doom. And that's all of them.

* * *

Sure: there's a limit to growth. But with all of space to play with I'll be happy to leave the problem for my descendants of 10,000 years hence to worry about.

I can hear the critics spluttering now. "But-but-but—what does this madman think he's *doing?* Flinging numbers like that around! Bringing in *asteroids.* It's absurd!"

Really? Remember, we haul quantities like that around here on Earth even now; in trains, and on boats, and it takes *far* more energy to process them here than it would in space. To get that energy here we must burn fossil fuels —which are really far too valuable as chemicals to set a match to—and put up with the resulting pollution. And, after all, I've assumed that we're going to supply the whole world with metals at the rate that we produce them from all sources—including recycling—here at ground level US of A. What's so absurd about it?

No, we won't be operating in space on this scale for a few years; but then we weren't producing all those tons of steel in 1930 either. Even the worst crunch models will not kill us off before 2020—a year in which we might very well be able to move asteroids around, boil them up for processing, and bring the resulting metals down for use here on Earth. It's a year in which we *certainly* will have Solar Power Satellites, always assuming that we *want* SPS. And there's approximately as much time between now and 1930 as now and 2020.

Yes. We live on a finite Earth. But there's a whole solar system out there. If we like we do not live on "Only One Earth." If we like, we live in a system of nine planets, 36 moons, a million asteroids, a billion comets, and a very large thermonuclear reactor/radiation source. It's all waiting for us out there. We've only to lift our heads out of the muck to find not only survival, but survival with style.

A Blueprint for Survival

This may be a unique century in many ways. In one respect it certainly is: this is the first time that mankind has had the resources to leave Earth and make his home in the solar system. No one doubts that we can do it. It takes only determination and investment.

Alas, we may be unique in another way: ours may be the *only* century in all of history when Mankind can break free of Earth. Our opportunity may not come again, *per omnia seculae seculorum*. Thus it could be that we have it in our power to condemn our descendants to imprisonment forever.

After the publication of "Survival with Style," a reader commented as follows: "I remain skeptical. By the time man is forced to accept population control, the world is going to be in a sadder state than it is now. And I doubt if nations will give up their armaments and their free school lunches in order to get the resources to mine the asteroids until the situation is so bad that we probably can't mine the asteroids in time to save us."

Unfortunately he may be right. There are no end of foreseeable crises, and enough of them could so deplete our resource base and technological ability that when we realize that we *must* go to space, we won't be able to get there. Furthermore, anti-technological sentiments are no joke; a great number of influential intellectuals have embraced Zero-Growth, condemn technology, and seem to want the next generation to atone for the sins of our forefathers. They do not appear to want themselves to atone; I haven't seen many leading intellectuals giving up their own luxuries, much less necessities, in order to make amends for the "rape of the Earth," "eco-doom," and the rest of what engineers and technologists are accused of. *We* shall continue to enjoy; but after us, The Deluge. Our children shall pay.

And of course if Zero-Growth has its way, our children *will* pay; but ours won't pay as much as the children of the people in the developing countries. Those kids are doomed with no chance at all.

Do not misunderstand. Were Earth our only source of energy and resources, I should probably myself be crying Doom. As it is, I fully support many conservation measures —and in fact I was writing pro-conservation articles as early as 1957. I've no use for wasters of Earth's bounty. But I've less use for those who would condemn most of the world to eternal poverty when there are ways that we can do something about it.

Incidentally, the Club of Rome, which sponsored the original computer studies leading to THE LIMITS TO GROWTH, and provides much of the intellectual fuel for Zero-Growth, has now sponsored a second report entitled MANKIND AT THE TURNING POINT (MATTP). This book, unlike LIMITS, is supposed to hold out some hope for the poor. By looking at the world as a set of 10 "regions" we can, say the authors of MATTP, divide the

49

wealth and sustain what they call "balanced growth."

Unfortunately they never tell us how. As one reviewer put it, "I do not find any clear explanation of the ways in which balancing out the regions of the world would lead to any lessening of the total demands of human civilization on the planet's living-space, resources, and vital eco-systems." (Frank Hopkins, in the October 1975 *Futurist.*) Moreover, the MATTP plan demands foreign aid at the rate of $500 billion *a year* at the end of a 50 year development period. True, there are plans with less massive foreign aid investments; but all are truly enormous.

And this is nonsense. No politician is going to run on a platform of international bounty. No democratic—or communist—nation is going to shell out wealth at that rate. And even if, by some miracle, the western nations were to divvy up with everyone else, the Second Report can't challenge one feature of THE LIMITS TO GROWTH: no matter how wildly successful we are in imposing Zero-Growth and population control, in 400 years the game will be over. We will have run out of non-renewable resources. Mankind will have no choice but to give up high-energy civilization and return to some kind of pastoral society.

Surely this is not a desirable goal? There may be those who dream of the simple life (and a lesser number who will actually choose to live it), but surely only a madman would impose it on everyone else without dire necessity? If there is any alternative, must we not take it?

* * *

There are alternatives. They aren't even very expensive compared to the MATTP plan. Take, for example, the detailed plans of Princeton professor Gerard K. O'Neill.

Details of what have come to be called O'Neill Colonies were first widely published in the September 1974 issue of *Physics Today.* The plan has been modified somewhat since that time, most recently by a week-long NASA sponsored conference of some of the biggest names in space exploration, but the basic concept remains the same: building self-sustaining colonies in space. O'Neill colonies have a major advantage: they are not only self-sustaining, but will be capable of building *more* colonies without further investment from Earth. Moreover, they will be able to make some important contributions to Earth's economy.

There's been a great deal of excitement in the science community, and of course among science fiction fans, although oddly enough most SF writers haven't put much about O'Neill colonies into print. I haven't because I assumed others would, and I was waiting for new details. Certainly much of the SF community is aware of the O'Neill concept. "Life In Space" is now a regular program item at science fiction conventions.

The basic O'Neill plan is for colonies able to support from 10 to 50 thousand people each. They will be located in the L4 and L5 points of the Earth-Moon system. Since not all readers will know what that means, and the location is important to the economics of the concept, let me take a moment to explain Trojan Points.

The equations of gravitational attraction are so complex that we can't really predict where planets, satellites, moons, etc., will be after long periods of time. Given high-speed computers we can make approximations, but we can't precisely solve problems involving three or more bodies except in special cases. A long time ago LaGrange discovered one of those special cases, namely, that when a system consists of three objects, one extremely massive with respect to the rest, and a third very small with respect to the other two, there are five points of stability: that is, things that get to those points tend to say there. These are often called "LaGrangian Points," and designated by the numbers L1, L2, ... L5. They are illustrated in Figure 8.

Of the five, three are not really stable; that is, if an object is perturbed out of L1, L2, or L3, it won't tend to return. The other two, L4 and L5, are dynamically stable, and it takes a special effort to get out of those locations. Left to themselves things put into points L4 and L5 will be there forever.

Figure 8

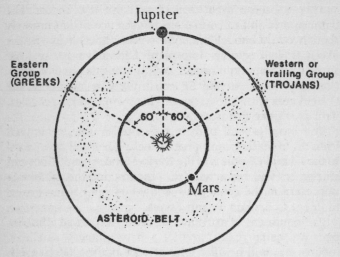

The Original "Trojan" Points

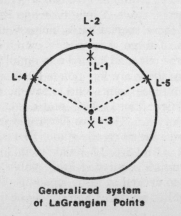

Generalized system
of LaGrangian Points

Nothing is left to itself, of course: there are more than three bodies in the solar system. Even so, satellites placed at those points would be stable over geological eras.

Points L4 and L5 are named *Trojan Points* because in the Sol-Jupiter system these points are occupied by a number of asteroids named after Trojan War heroes. The Trojans trail Jupiter, while the Greeks lead. Unfortunately the custom of naming the Eastern group for Greeks and the Western for Trojans wasn't established before one asteroid in each cluster was named for the wrong class of hero; thus there's a Trojan spy among the Greeks, and vice-versa.

Because of the perturbing influence of other planets, Trojan Points aren't really "points"; the Trojan asteroids drift around within a sausage-shaped area about an AU (93,000,000 miles) in diameter, while objects in the Earth-Moon Trojan Points would tend to drift a few thousand miles one way or another; but they're stable *enough* for our needs. Colonies, and supplies for the colonies, once arrived at L4 and L5, won't go anywhere. The points are, of course, 240 thousand miles from Earth, and an equal distance from the Moon.

O'Neill colonies will be *big*. Even the first model, which is intended as an assembly base and factory, will be several kilometers in diameter. Later models will be bigger. One design calls for a cylinder 6 kilometers in diameter and several times that in length. The cylinders slowly rotate to provide artificial gravity. The exact gravity wanted isn't known yet, but it will certainly be less than that of Earth, possibly low enough that man-powered flight (yes, I mean people with artifical wings) will be not only feasible, but the usual means of personal transportation. As O'Neill points out, a great number of energy-consuming activities required for civilized life on Earth won't be needed in the colonies. Roads and automobiles and trucks aren't wanted.

It's possible to wax poetic about the idyllic life in O'Neill colonies, but I won't do that. In the first place I may be far-out technologically, but I don't think people are likely to live in Utopian style no matter how pleasant their environment. The important point is that life can be pleasant, and certainly possible, in space colonies.

These colonies are to be self-sufficient: they have more than enough agricultural space to feed their inhabitants. More importantly, they have a product to sell Earth: energy. It is perfectly feasible to collect solar power, convert that to electricity, and beam the juice down to Earth by microwave. Tests show the cycle, DC to DC, is about 65% efficient—and of course most of the wasted energy doesn't get to Earth, but stays in space. There are a number of designs for the Earth-based receivers. The one I like best is a grid of wires several meters above ground: energy densities are low enough to allow cattle to graze comfortably in the pasture below. Alternatively, you could plant orange groves there.

All this sounds lovely, but surely it's a bit far-fetched? No. O'Neill's designs use present technology. There are no super-strong materials and no magic systems. We could begin building an O'Neill colony this year, occupy it in 1990, and by the year 2000 have a couple more of them built, in which case we could also be supplying about as much power to Earth as the Alaskan pipeline will. In 20 more years space could supply nearly all the US electric power.*

So why don't we do it?

It's bloody expensive, that's why. Make no mistake: this would be a costly undertaking, on a level of effort comparable to the Interstate Highway System, or the Viet Nam War.

It would not, in my judgment, be nearly so expensive as Zero-Growth, but unfortunately the costs of space colonies are *visible*. They're direct expenditures. The costs of Zero-

---

*Readers with more interest in O'Neill colonies should send $20 to the L5 Society, 1620 N. Park, Tucson AZ 85719, which publishes a newsletter and lobbies for NASA support for space colonies. I thoroughly recommend their organization.

Growth are hidden, since the most costly part is in potential not used and goods not created.

In the December 5, 1975 *Science* (the prestigious publication of the American Association for the Advancement of Science) Dr. O'Neill presents an economic analysis of satellite solar power stations (SSPS's) and Space Manufacturing Facilities. He comes up with total costs ranging from a low of $31 billion—about the proportion of GNP that Apollo cost—to a high of $185 billion. He also discusses benefits from the electric power produced by SSPS's, and concludes that over a 40 year period the facilities would show actual profits from sales of power alone.

As a co-discoverer (with Poul Anderson) of what was once known in the aerospace industry as Pournelle's Law of Costs and Schedules ("Everything takes longer and costs more"), I tend to distrust Dr. O'Neill's numbers. It hardly makes any difference. The important point is that the program is feasible. We could afford it. Take a worst-case. Suppose it takes 25 years, and the total cost is 50 Apollo programs, that is, a round one trillion bucks. The money must be spent at $40 billion a year for the next 25 years, which comes to $200 a year for every man, woman, and child in the US. In my own family it would be about $1000 a year.

That's a lot of money. Worth it, I think; the benefits are literally incalculable. For example, by the year 2000 the US will need 2 billion tons of coal annually simply to operate our electric power system. Nuclear power plants could reduce that substantially, but the nuclear industry is in deep legal—not technological—trouble. It would be worth a lot to me simply to avoid the strip mines that 2 billion tons a year will require.

Moreover, the space budget isn't going to be simply tacked onto the national budget. All of the money will be spent here on Earth—people living in Lunar and space colonies have no need for Earth dollars, and what they physically import is tiny compared to the salaries that will be paid to Earth workers manufacturing products for the col-

Figure 9

BE OF GOOD CHEER, THERE'S HOPE FOR US YET. . . .

O'NEILL COLONIES

Assume each colony can build one new colony in ten years.
Assume there are only 50,000 people in each colony.
Assume we don't get started until 2020 AD . . .

By 2283 AD (Standing Room Only at 4% population growth),
There will be
3.85×10^{15} people
living in O'Neill Colonies.

Comparing favorably to the 6.8×10^{14} people required to cover the Earth. . . .

(WELL, we never said we'd give the Club of Rome a monopoly on exponential curves.)

ony program. With $40 billion a year in high-technology industries, we can eliminate a number of "pump-priming" expenditures and dismantle several welfare and unemployment compensation schemes as well.

Of course we won't really need to spend that kind of money, and I suspect we can start getting returns on that investment before 25 years. O'Neill himself thinks in terms of some $5 billion a year, which works out to $25 a head for each person in the US; and the colonies have got to be worth *that* if only in entertainment value.

Now how can something as complex as space colonies be built for that low a price? And wouldn't it be cheaper to build space manufacturing facilities in near-Earth orbit rather than going out to L5?

That's the beauty of the O'Neill concept. All the building materials for the colonies must, of course, be put into orbit —but they need not come from Earth. Most of the raw materials for the L5 colony will come from the Moon.

The Moon has one twentieth the gravity well that Earth does. The colonies will be in stable Trojan Points. Put those two data together and you reach an interesting conclusion: much of the mass of the colonies need never have been launched by rockets at all.

There are several devices for getting lunar materials to the L5 point. One involves a simple centrifugal arm: a big solar-powered gizmo similar to the thing used to pitch baseballs for batting practice. It flings gup, such as unrefined Lunar ore (25-35% metal, from our random samples) out to the L5 point, and the laws of gravity keep it there. Refining takes place at the colony, and the slag is useful as dirt, cosmic ray shielding, and just plain mass. There's also oxygen in them there rocks.

Another workable device is the linear accelerator or mass driver—a long electric sled as used in countless science fiction stories. Both these can be built with present technology.

Obviously, then, O'Neill colonies have a prerequisite,

namely, a permanent Lunar Colony. Now that's certainly within present-day technology; I once did studies that demonstrated that with technology available in the 60's we could keep astronauts and scientists alive for years on the Lunar surface, and things have come a long way since then.

The Lunar Colony will need at least one near-Earth manned space station, since Earth-orbit to Lunar-orbit is the most efficient way to transport large masses of materials from here to there. The Lunar Shuttle will be assembled in space, and won't have all that waste structure that would enable it to withstand planetary gravity; thus it can carry far more payload per trip.

It's here that I think the profits come in. Skylab demonstrated that space manufacturing operations have fantastic potential profits. There are things we can make in space that simply cannot be made on Earth. Materials research benefits alone might pay for the space station. Certainly the potential for Earth-watch operations, pollution monitoring, better weather prediction, increased communications, and all the other benefits we've already got from space, will contribute to profits as well.

And once space shows a visible return on investment, we may well be on our way.

So. The prerequisite for the space station is the Shuttle; and there's the weak point. The Shuttle is in trouble. There are a number of Congresscritters who'd like nothing better than to convert the Shuttle into benefits for their own districts. There are plenty of intellectuals who continually do cry "Why must we waste money in space when there are so many needs on Earth?" The obvious reply, that most of our expenditures on Earth seem to have vanished with no visible benefit, while our space program has already just about paid for itself in better weather prediction alone, does not impress them.

There are also the Zero-Growth theorists who see investment in space not as a mere waste of money, but as a positive evil.

We are close to breakthrough. For a whack of a lot less than we spend on liquor, or on cigarettes and cosmetics, on new highways we don't need, on countless tiny drains that fritter away the hopes of mankind, the United States alone could break out of Earth's prison and send men to space. The effect on future generations is literally incalculable. We *can* do it; but will we?

* * *

I wish I were sure that we would; or that if we of the US don't do it, somebody else will; but I am not. There are just too many disaster scenarios. A Great Depression. War. The triumph of anti-technological ideology. The continued ruin of our educational system—in California, with 30 State Universities, there is not one in which bonehead English is not the largest single class, and the retreat from excellence (called democratization and equality of opportunity) races onward. Any of these, or all of them at once, could throw away an opportunity that may never again come to mankind.

So what can we do?

For one thing, we can organize at least as well as the opposition. Science fiction readers may have mixed emotions about "ecology" movements, consumerism, Zero-Growth, and the like, but I think we have not lost our sense of wonder, nor abandoned our hopes. We have not given up the vision of man's vast future among the stars. We have not traded the future of man for a few luxuries in our time.

Unfortunately, we have no voice, or rather, we have a myriad of voices, none very effective.

In the 50's a number of us in the aircraft industry used to bootleg space research. There wasn't any budget for that crazy Buck Rogers stuff. Most of us believed we would see the day when the first man set foot on the Moon. We didn't believe we'd see the last one. I hope we haven't.

Like many of us who recall pre-Sputnik days, I alternate between hope and depression. Recently I have seen one hopeful sign, although it is a bit frightening.

As I write this it appears that the Soviets have built lasers sufficiently powerful to blind our infra-red observation satellites. These satellites are in very high orbits, meaning that the Soviet lasers must be extremely powerful. One old friend who has remained in the industry told me at a New Year's party that the Soviets must be at least 5 years ahead of us, and this in a field in which we thought we were supreme.

Why is this hopeful? Isn't it rather frightening?

It's frightening if you think the Soviet Union may fall or be under the control of convinced ideologists willing to trade part of their country for all of the world. There is nothing in Marxist ideology to forbid that—indeed, any communist who has the opportunity to eliminate the West and thus bring about the world revolution, and who fails to do it because of the price in human lives, is guilty of bourgeois sentimentality. So yes, it's frightening that the Soviets may have taken several long strides toward laser defense against ICBM's.

It's hopeful, though, in that it may stimulate us to get moving in large laser R&D. In my judgment, defense technology is the ideal way to conduct an arms race, if you must have an arms race. (And it takes only one party to start a race, unfortunately.) Defense systems don't threaten the opponent's civilian population. They merely complicate offensive operations, hopefully to the point where no sane person would launch an attack; and they give some hope that part of your own civilian population may survive if worst comes to worst.

If we can't justify space operations in terms of benefits to mankind, then perhaps we can sell them as defense systems? Big lasers can be used as space launching systems. If built they can put a good bit of material into orbit, thus making the manned space factory economically feasible and nearly inevitable; and once in Earth orbit, as I said in the first of these columns exactly two years ago, you're halfway to anywhere.

Specifically, we'd be halfway to an era of plenty without pollution; halfway to assuring that our descendants won't curse our memory for throwing away mankind's hope for the stars.

How Long to Doomsday?

"WHILE YOU ARE READING THESE WORDS FOUR PEOPLE WILL HAVE DIED FROM STARVATION. MOST OF THEM CHILDREN."

Thus opens Paul Ehrlich's THE POPULATION BOMB.

"It seems to me, then, that by 2000 AD or possibly earlier, man's social structure will have utterly collapsed, and that in the chaos that will result as many as three billion people will die. Nor is there likely to be a chance of recovery thereafter. . . ."

Thus closes a popular article by Dr. Isaac Asimov, perhaps the best-known science writer in America.

It would not be hard to multiply examples of doom-crying among science fiction writers, or, for that matter, the American intelligentsia. There are dozens of stories and articles describing life in these United States after the year 2000 as poor, nasty, brutish, and short—although hardly solitary as Hobbes would have it.

Much of this doomsaying springs from three original sources which are endlessly requoted: Ehrlich's work previously mentioned, and two outputs from MIT: WORLD DYNAMICS and THE LIMITS TO GROWTH. All are essentially mathematical trend projectons, with the MIT studies employing complex and highly detailed computer models.

Strangely, intellectuals including science fiction writers have a lot of confidence in these economic models, although they have very little in the ability of social or physical scientists to save us. It's almost impossible to overestimate the influence of these three books. Writers make predictions based upon them; teachers quote them endlessly, or worse, quote secondary and tertiary sources which draw their ideas from them.

The result is that these works and the view they represent have become "conventional wisdom" for the young. DOOM is "in the air" so to speak; a great part of our younger generation is convinced that no matter what we do, no matter how much we discover or learn, we are finally and inevitably doomed. If Isaac Asimov says we are finished, then what hope have we?

Even when the result is not total defeatism—after all, if we're doomed no matter what we do, why not "tune in, turn on, and drop out"?—the influence of this view is crucial. For millennia the concept of progress has been the driving force of Western civilization. Our philosophy was simple: hard work and study would save us. "Ye shall know the truth, and the truth shall make you free." Now all that is lost. Western civilization has lost faith in progress. Our only salvation, a new era of intellectuals say, is through Zero-Growth; "Small is Beautiful"; "soft energies"; and the like. Political figures including the Governor of California and the President of the United States base their future planning on this philosophy; they have specifically abandoned the older idea that "knowledge is power" and that good research and technology development will bring about an era of plenty.

Yet—are we doomed? Surely the works which generated that view deserve analysis.

* * *

First: the blurb that opens Ehrlich's book is clearly wrong. My copy was published in 1969, a year in which about 53 million people died from all causes. It takes four seconds to read the blurb, so for one person to die each second, 31.5 million—about 60 percent of all deaths—would have had to be from starvation.

Taking the UN cause-of-death statistics and being as fair as possible by including as "starvation" any cause related to nutrition—diphtheria, typhus, parasitic diseases, etc.—we get about a million, or some 5 1/2 percent. Dr. Ehrlich is off by a factor of ten.

Actually, world agriculture is keeping up with population. At the Mexico City meeting of the American Association for the Advancement of Science in 1975, Dr. H. A. B. Parpia, the senior professional of the UN's Food And Agricultural Organization, told me that just about every country raises more than enough food to be self-sufficient. The food is grown, but sometimes not harvested; or if harvested, spoils before it can be eaten. In many countries vermin get more of the crop than the people: insects outeat people almost everywhere. The pity is that the technology to harvest and preserve enough for everyone exists right now.

Now this essay is not intended to be a Pollyanna exercise. There's no excuse for relaxing and saying that hunger is a myth. It isn't. But simple food storage technologies, and research into non-damaging pesticides and pest control methodologies, could stop famine in most of those parts of the world where that horseman still stalks the land. Other simple technologies—even mylar linings for traditional dung-smeared grain storage pits—would save lives.

We know how to do it; but we won't unless we're willing to try. We won't get anywhere sitting around crying

"Doom!"

Yet according to Dr. Ehrlich's book, "The battle to feed all of humanity is over. In the 1970's the world will undergo famines—hundreds of millions of people are going to starve to death in spite of any crash programs embarked upon now."

Fortunately that didn't happen; but the doomsayer viewpoint, which did not stop agro-engineers from making efforts despite the flat prediction that their efforts were useless, did invade our schools so successfully that a new generation of students believes in Doom as thoroughly as ever did a Crusader in the holiness of his cause.

* * *

The other side of the coin was expressed in the Hudson Institute's THE YEAR 2000, which points out that the level of rice yield per acre in India has not yet equalled what the Japanese could do in the Twelfth Century. Another analyst, Colin Clark, has shown that if the Indian farmer could reach the production levels of the South Italian peasant, there would be no danger of starvation in India for a good time to come.

In other words, it doesn't even take Miracle Rice, fertilizers, and a high-energy civilization to hold off utter disaster in the developing countries. It only takes adding technology to traditional peasant skills—indeed, the kind of thing advocated by E.F. Schumacher in his SMALL IS BEAUTIFUL-ECONOMICS AS IF PEOPLE MATTERED. Showing people how to use mylar and simple non-persistent fungicides for food storage along with peasant agricultural methodology will hold the line against famine—for a while.

Moreover, we *have* new technologies. There *are* means for increasing protein production. More protein in childhood would cut back infant diseases like kwashiorkor and "red baby"; those diseases have the effect of permanently lowering adult IQ by about 20 points. What if the next generation of a developing country were "20 IQ points" more intelligent? For many of the ignorant of the world are not stupid; but they may be *stunted*.

But the doomsters have an answer. If we help those people feed themselves, they'll only breed to famine again. Worse, they'll demand industry. They'll strip-mine phosphates and poison the seas (as shown by Cousteau on a recent film). What's the point of helping them? Doom is still around the corner.

The best answer is that historically people haven't done it. When nations reach a high level of technology—and of infant survival—the fertility rate falls. The US appeared to be an exception to that with the WW II "baby boom," but now that squiggle in the fertility rate has passed. The girls born in 1944 are 35 now, leaving their child-bearing period, and the number of girls born per fertile girl in the US has fallen to an all-time low: so low that now one occasionally hears economists advocate bonuses for larger families! The same is true of the other industrialized nations. Populations of wealthy nations do *not* rise without limit.

Yet—in our schools and colleges and universities straight unadulterated Malthusianism is taught and learned and has become "conventional wisdom."

* * *

There's another form of doom not so fashionably discussed: the Marching Morons (that is, the least successful tend to have the most children). It's a problem we must face; but it's doubtful that before the year 2000, or even 2500, it will have destroyed our social institutions.

As a matter of fact, given present population trends, the US won't have very many more people in 2000 than now. Population *is* growing, albeit slowly: there's a "bow wave" generated by the World War II "baby boom," and of course there is always immigration—both legal and illegal. Still, best projections show us peaking in about 2025 with population then declining to its present level—where it will stay.

Suppose that never happens, and we reach 350 million people before something stops US population growth. The area of the United States is about 9.5 million square kilometers; of that, some is water, and some simply uninhabitable. Call it 8 million even, and we have a present population of about 26 people per square kilometer.

If we reach 350 million people—and few projections show us getting there in 50 or even 100 years—we would have 43.5 people/km², a big increase. Some writers say that we will be driven stark, staring mad by overcrowding, and this well before 2020 AD; Asimov, recall, expects Doom before the year 2000, primarily from this cause.

We'll be inundated with personal contacts, at each other's throats, sleeping in hallways and abandoned automobiles; mothers will kill and eat their children; few will have any incentive to work; all except the *very* rich will be in utter misery as civilization collapses.

Well, what civilized countries have population densities higher than our doom-level of 43.5 that we might reach in 50 years?

Practically all of them. West Germany, a not uncivilized place, has 244 people/km², equivalent to 1.9 *billion* people in the US! Denmark has 114 people/km²; France 93; England and Wales, 322. Even Scotland, with its highlands and islands and hills and moors has 66.

What densities can people stand and remain sane? No one really has an answer to that. But the Netherlands, a charming place, has 319 people/km²; the Channel Islands has 641; and Monaco, the densest place on Earth, has 16,000!

Of course the U.S. could not be packed like Monaco or England. We would not like it if our country were as thickly populated as Denmark (although our eastern seaboard is more densely populated in some places right now); but surely we would not all go insane if we lived as close together as the Scots!

And note well—Europe has *always* been the most densely populated area of the Earth; far more so than Latin America or Asia. Latin America, in fact, is almost under-populated compared to Europe.

No: our civilization will *not* collapse from overcrowding, at least not in the forseeable future; and the silly assertions about imminent DOOM from crowding come mostly from a failure to do elementary calculations. (Incidentally: although our schools abound with doomcrying teachers, there is a hopeful sign, namely, that pocket calculators are readily available and very cheap. What will happen when the nation is ruled by a generation that habitually uses elementary arithmetic because it's easy and one doesn't make mistakes? The effect could be highly beneficial.)

Moreover, we have the technology right now to support a large population while preserving wilderness. Soleri's *Arcologies* is a fascinating book: he shows enormous cities providing for millions built on a few square miles of land, leaving parks and woodlands around them.

We even have the technology to make the *whole Earth* a park if we really wanted to: by going to space for our messiest operations we could end most pollution on this planet; in a hundred years we could, if we'd just get to work, restore practically all of North America to her pristine state.

Less ambitiously, I have "designed" a city for a story about Los Angeles in the future: in my design, a 50-level building contains lodging, stores, conveniences, recreation, employment, and the transportation for 250,000 people. The "Independency of Todos Santos" is 2 miles on a side and sits on an area 4 miles on a side; 250,000 people in 16 square miles. Fewer than a hundred such buildings would hold the entire U.S. non-farm population—and my structure is not only small by Paulo Soleri's standards, but uses very little technology we don't already have.

When Larry and I began our story, incidentally, we thought it a bit far-fetched that people might prefer to live in our "city" rather than in suburbs. Now we've seen condominiums with full conveniences, recreation, transportation, even employment; they cost more than the suburbs, yet most of their inhabitants are refugees from suburbia. It no longer seems fantastic at all. Why not live in a convenient place where you can walk to work, take an escalator to the opera, and a train to the beach? Why fight commuter traffic?

* * *

The evidence is plain: the population bomb will not kill us, nor even drive us mad, within our lifetimes. Certainly we can't keep on doubling populations as fast as we have in the past—but why assume that we will? When the Reverend Thomas Malthus made his gloomy predictions, someone blindly running off the exponential growth equations (doomcryers are fascinated by exponential growth, although I don't know of a single case of it in nature) would have calculated that England in 1970 would have 400 million people instead of the present 55 million.

Population stability won't happen of itself, but most of the really alarming population growth has been through the prolonging of life. Birth rates have declined through this century, but people live longer despite wars, famines, pollution, insecticides, crowding, and all the other forms of doom. Since there's a limit to just how long anyone can live, the death rate is due to climb before 2000. Already many countries have aging populations; including the US of course. It was never true since Colonial times that "over half of the people are under 25" and it gets less true all the time. Much of the "population explosion" is a one-time artifact, and you can't simply apply equations of exponential growth to predict the future.

Certainly population pressure *can* finish us off; but why believe we'll get to the *Soylent Green* stage before something is done about it? The evidence is that the technologically advanced countries, and even some not so advanced, such as China, have already done something about it; and certainly we won't be destroyed by overpopulation before 2020 or even 2100.

* * *

If we have defused, or at least delayed, the population bomb, what's the next thing to kill us? Asimov says that if we survive going mad from overcrowding we'll still be finished because of energy limits.

We've dealt with this in another chapter; there are non-polluting systems to supply us with all the energy we need to run not only Western high-energy civilization, but to industrialize the world. Ocean Thermal; Solar Power Satellites; hydrogen fusion; any one would do the job provided we have the gumption to build it.

Another doom, the rising levels of carbon dioxide which convert our planet into a sterile hothouse, falls to quantitative analysis: it's true enough that the levels of CO_2 in our atmosphere have risen since 1900, but not so sharply as all that; and before they can get to a point where they do any real damage, we'll have run out of fossil fuels to burn. It's true we should concern ourselves with the climatic future of the planet—there's evidence that we're about due for an Ice Age—and some evidence that we'd be in one *now*, if it were not for all the fuels we've burned and the heat we've introduced. But that too is something we can deal with, provided we don't lose faith in ourselves.

In fact—on any careful analysis we're not doomed at all. Quite the opposite. We have it in our power to go to space; to liberate man from the prison of Earth; to get humanity spread across a number of planets and moons and space colonies so that no one disaster can exterminate us.

We can turn the Earth into a park.

This is the first generation in history to not only be concerned about ecology and conservation, but also to have the resources to do something practical about them without condemning much of the world to starvation. This generation can give Mankind the stars and planets.

We live in one of the most exciting times of all history. Surely we can do better than cry Doom!

That Buck Rogers Stuff

The young lady was very serious, and although I might have wished that she looked like an ogre with raucous voice and nose and chin meeting in front of her lips, she was actually very professional in appearance, highly attractive, and according to most objective standards, intelligent. My wife and I had come to a typical Los Angeles show-business party, and the young lady had been waiting for me. Before I could get properly into the room she advanced menacingly.

"You write science fiction," she accused. "Escapism. What good does it do to get people dreaming about that Buck Rogers Stuff?" (I swear it, she used that phrase, the same one that countless teachers used in the days of my youth when they caught me reading Astounding Science Fiction.)

Naturally, she had A Cause. "We spent billions for what? For some pieces of rock and pretty pictures on television!

And there are millions out of jobs, we need better schools, and—"

Readers have probably had similar experiences and can finish off the speech for themselves. It's not the only time I've been put to The Question: "Why throw money away on space when there's so much that needs doing here on Earth?" All right, let's talk about space and see just how far we can get.

First, for a really beautiful job of presenting what we've already got out of space, see the NASA SPINOFF documents; they print another each year, and they tell what new economic impact space research has had on American lives.

The SPINOFF documents are written by Neil Ruzic, who's also the author of an excellent book on the future uses of the Moon called WHERE THE WINDS SLEEP. Between the SPINOFF annuals and Ruzic's book you can find plenty of answers to the silly question about why we should spend money on space.

In fact, the problem is knowing where to begin. Weather predictions? Remember when the weatherman was a joke? Now true, the Weather Bureau makes some mistakes even yet; but not very many, and almost never when it comes to hurricanes. You can show that the space program has pretty well paid for itself just in better weather forecasting alone.

Those concerned about pollution will be pleased to hear that Earthwatch satellites finally give us a chance to see the real effects of pollution. Mining prospecting has been revolutionized by satellite photography. The international Food and Agricultural Organization in Rome can, from satellite data, get a good forecast of famine areas and global food production.

That's all satellite stuff. Industry benefits are nearly incalculable, and I don't mean frivolities like Teflon frying pans. Test procedures and quality control: the inspection methods developed for man-rating spacecraft and boosters

72

are now routinely used in building better plows, tractors, automobiles, skis, hiking boots and packframes, electronic equipment, and darned near anything else you can think of.

In my early days in the space program one of the hardest jobs we had was monitoring physiological conditions in a stress environment. Just getting an ordinary electro-cardiograph (EKG) through a pressure wall required great ingenuity. We invented a number of such devices; we had to. My own inventions are long since obsolete —but the space medicine technology that grew out of our early efforts is routinely used in hospitals and clinics all over the world. Mass spectrometers to analyze exhaled breath; microminiature EKG systems worn by hospital patients and displaying abnormalities to the duty nurse; blood analysis equipment; even heart condition diagnosis from moving vehicles; all routine, and all developed as part of the NASA package.

Your tires last longer, you can buy large fiberglass structures, firemen can keep your house from burning, your electrical system is simpler, crash helmets work better (remind me sometime to tell you about the purchase order for "nine freshly-killed human male corpses, ages 21 to 40 at time of death, must not have any abnormalities of brain or upper spine; expendable research item; no salvage value." The Purchasing Officer's reaction to that was, uh, interesting); driver-training simulators work, paint lasts longer, and golf clubs do a better job of driving the ball.

"Whoa. That's all technology, and technology is evil. It causes pollution, and kills people in wars, and—"

And at that point my usual reaction is a loud "Aaargh!" and a burning desire to find a drink. Quickly. Especially when it was said by a young person wearing a thin wristwatch and polyester imitations of honest blue denim, driving a Mercedes, and feeling committed because he hasn't eaten table grapes for *weeks*. I should control that reaction, of course; but if I were able to do that I'd probably still be in aerospace management instead of living the unnatural life of a writer.

Still, such people ought to be answered; our whole future may depend on it. Let's try.

California's Governor Jerry Brown has built himself quite a reputation by pushing "Alternate Technology" and the philosophy that goes with it. "Make do. Expect less. Conserve. Smaller is better. Recycle. Be satisfied with what you have. There's Only One Earth."

Now there are some attractive points about all that. Moreover, the vision of a stable, low-to-zero-growth economy, concentrating on adventures of the mind, with a lot of "cottage industry" can be a noble one. It's probably possible, too—for us, and for a while.

It is not a philosophy likely to appeal to the poor of this world. Like it or not, a conservation-oriented low-growth world economy dooms most of the world's people to wretched poverty. But what has that to do with *us?* Can we not, ourselves, change our ways and let others go theirs?

Probably not. Like it or not, we've got most of the technology—and we don't have enough to develop the Earth to a point of satiation. If all the world gets rich through the same wasteful processes we employed, we're probably in big trouble. Worse, what of our grandchildren? The Earth's resources will not last forever; and what then?

I've argued here before that this generation is crucial: we have the resources to get mankind off this planet. If we don't do it, we may soon be facing a world of 15 billion people and more, a world in which it's all we can do to stay alive; a world without the investment resources to go into space and get rich. Usually I think it won't come to that; it's only in odd moments—such as when faced with The Question—that I get depressed.

I don't think it will come to that, because the vision of the future is so clear to me.

We need realize only one thing: we do not inhabit "Only

One Earth."

Mankind doesn't live on Earth. Man lives in a solar system of nine planets, 34 moons, and over half a million asteroids. That system circles a rather small and unimportant star that is part of a galaxy containing tens of billions of stars. Only One Earth, indeed! There are millions of Earths out there, and if we use up this one, we'll just have to go find another, that's all.

We needn't use up this one. In a previous chapter I went through the numbers: how we can, with present-day technology, deliver here to Earth as much metal for each person in the world as the US disposed of per capita in the 60's. We can do that without polluting our planet at all, and we can keep it up for tens of thousands of years. The metal is out there in the asteroid belt. For starters we don't even have to look very hard; most of the asteroids were once spherical, large enough to have metallic cores, and now the worthless gubbage topside has been knocked away, exposing all that lovely iron and lead and tin and such we'll need to give the wretched of the Earth *real* freedom.

Why not? The refinery power's there; the Sun gives it off for free. We have a propulsion system to get us to the asteroids; Project NERVA was cancelled, but the research was done, and it wouldn't be that hard to start up again. Nuclear-powered rockets would be rather simple to build, if we wanted them.

But first we'll need a Moonbase. We can get that the hard way, carrying stuff up bit by bit from the top of disintegrating totem poles, but there are easier ways.

We could do it in one whack. Project ORION was also cancelled, but we could build old Bang-Bang in a very few years if we wanted to. ORION used the simplest and most efficient method of nuclear propulsion of all: take a BIG plate, quite thick and hard; attach by shock-absorbers a large space-going capsule to it; put underneath one each atomic bomb; and fire away.

Believe me, your ship will move. When you've used up

the momentum imparted by the first bomb, fling another down underneath. Repeat as required. For the expenditure of a small part of the world s nuclear weapon stockpile you have put several *million* pounds into orbit, or on the Lunar surface.

But that will cause fallout.

Yes; some. Not very much, compared to what we have already added to background radiation, but perhaps enough that we don't want to use ORION—although, he said happily, ORION is one reason why I think we'll eventually do what has to be done, even if this generation fails in its duties to the future. ORION is cheap and the bombs won't go away; if we're still alive in that grim world of 15-20 billion and no space program, *somebody's* going to revive Bang-Bang and get out there.

ORION gets a few big payloads to orbit or the Moon. A more systematic way would be to build a big laser launching system and make it accessible to anyone with a payload to put into orbit. Freeman Dyson calls laser launch systems "space highways." The government builds the launch system, and can use it for its own purposes; but it also gives private citizens, consortiums, firms, a means of reaching orbit.

Dyson envisions a time when individual families can buy a space capsule and, once Out There, do as they like: settle on the Moon, stay in orbit, go find an asteroid; whatever. It will be a while before we can build cheap, self-contained space capsules operable by the likes of you and me; but it may not be anywhere near as long as you think.

The problem is the engines, of course; there's nothing else in the space home economy that couldn't, at least in theory, be built for about the cost of a family home, car, and recreational vehicle. But then most land-based prefabricated homes don't have their own motive power either; they have to hire a truck for towing.

It could make quite a picture: a train of space capsules departing Earth orbit for Ceres and points outward, towed by a ship something like the one I described in "Tinker." Not quite Ward Bond in "Wagon Train," but it still could make a good TV series. The capsules don't have to be totally self-sufficient, of course. It's easy enough to imagine way stations along the route, the space equivalent of filling stations in various orbits.

Dyson is fond of saying that the US wasn't settled by a big government settlement program, but by individuals and families who often had little more than courage and determination when they started. Perhaps that dream of the ultimate in freedom is too visionary; but if so, it isn't because the technology won't exist.

However we build our Moonbase, it's a very short step from there to asteroid mines. Obviously the Moon is in Earth orbit; with the shallow Lunar gravity well it's no trick at all to get away from the Moon, and Earth orbit is halfway to anywhere in the solar system. We don't know what minerals will be available on the Moon. Probably it will take a while before it gets too expensive to dig them up, but as soon as it does, the Lunatics themselves will want to go mine the asteroids.

There's probably more water ice in the Belt than there is on Luna, so for starters there will be water prospectors moving about among the asteroids. The same technology that sends water to Luna will send metals to Earth orbit. I've already described one ship that can do the job. There are others. The boron fusion-fission process is a good example.

Take boron-11 ($^{11}B_5$). Bombard with protons. The result is a complex reaction that ends with helium and no nuclear particles. It could be a direct spacedrive. For those interested, the basic equation is

$$^{11}B_5 + p = 3(^4He_2) + 16MeV$$

and 16 million electron volts gives pretty energetic helium. The exhaust velocity is better than 10,000 kilometers/second, giving a theoretical specific impulse (I_{sp}) of something over a million. For comparison the I_{sp}

of our best chemical rockets is about 400, and NERVA manages something like 1200. The boron drive needn't be used very efficiently to send ships all over the solar system.

Meanwhile, NERVA or the fission-ion drive I described in an earlier column will do the job. In fact, it's as simple to get refined metals from the asteroid belt to near-Earth orbit as it is to bring them down from the Lunar surface. It takes longer, but who cares? If I can promise GM steel at less than they're now paying, they'll be glad to sign a "futures" contract, payment on delivery.

It's going to be colorful out in the Belt, with huge mirrors boiling out chunks from mile-round rocks, big refinery ships moving from rock to rock; mining towns, boom-towns, and probably travelling entertainment vessels. Perhaps a few scenes from the wild west? "Claim jumpers! Grab your rifle—"

Thus from the first Moonbase we'll move rapidly, first to establish other Moon colonies (the Moon's a *big* place) and out to the asteroid belt. After that we'll have fundamental decisions to make.

We can either build O'Neill colonies or stay with planets and Moons. I suspect we'll do both. While one group starts constructing flying city-states at the Earth-Moon Trojan Points, another will decide to make do with Mars.

Mars and Venus aren't terribly comfortable places; in fact, you probably won't want to land on Venus at all until it has been terraformed. Between Mars and Venus, Venus is the easier to make into a shirt-sleeves-inhabitable world. It requires only biological packages and some fertilizers and nutrients, and can be done from Moonbase or, in a pinch, from Earth itself. Still, though Venus may be the simpler job, Mars is likely to come first, simply because you can live there before terraforming; there will be dome colonies on the Red Planet.

I wrote a story ("Birth of Fire") describing one Mars-terraforming project: melt the polar caps and activate a number of Martian volcanoes to get an atmosphere built up. Isaac Asimov described the final step many years ago: get your ice from Out There, at Jupiter or Saturn, and fling it downhill to Mars. Freeman Dyson points out that there's enough ice on Enceladus (a Saturnian moon) to keep the Martian climate warm for 10,000 years. The deserts of Mars can become gardens in less than a century.

Dyson's scheme didn't even involve human activity on Enceladus; robots and modern computers could probably accomplish the job. They've only to construct some big catapults on the surface of Enceladus, and build some solar sails. Dyson suggests robots because the project as described would take a long time, and human supervisors might not care for the work; but I suspect we could get plenty of volunteers if we needed them. Why not? No one could complain that the work was trivial, and you couldn't ask for an apartment with a better view than Saturn's Rings!

Moonbases. Lunar cities. Mining communities in the asteroid belt. Domed colonies on Mars, with prospects for terraforming the planet and turning it into a paradise. An advanced engineering project headquarters on Enceladus. Pollution controlled on Earth, because most polluting activities would go on in space. Near-Earth space factories. Several to hundreds of city-states at the Trojan Points of the Earth-Moon system. A space population of millions, with manned and unmanned ships stitching all the space habitats together. This is not a dream world; this is a world we could make in a hundred years!

In 1872 a number of Kiowa and Comanche chiefs were taken to Washington by Quakers in an attempt to show the Indians just what they were facing. When they returned to talk about the huge cities, and "a stone tipi so large that all the Kiowa could sit under it," they were not believed. One suspects that if the Quaker schoolmasters had been magically transported to the Washington of 1979 and then returned to their own time, they would not be believed either. A nation of over 200 million people? Millions of tons of concrete poured into gigantic highways? Aircraft larger than the biggest sailing ships? City streets brightly lit at night? Millions of tons of steel, farmlands from Kansas to California . . .

Building a space civilization in the next hundred years will be simpler than getting where we are from 1879. We already know how to do it. We probably don't know how we *will* do it; certainly the very act of space exploration will generate new ideas and techniques as alien to us as nuclear energy would have been to Lord Rutherford or Benjamin Franklin; but we already know how we *could* do it. No basic new discoveries necessary.

In the 1940's I did a class report on space travel. I drew heavily from Astounding, from Heinlein's Future History, from Willy Ley's books on rockets and space travel (and certainly never thought I would someday be science columnist in the same magazine as Willy). My teachers were tolerant. They let me do it. They didn't believe in suppressing their pupils. Afterwards, though, the physics teacher called me in for a conference: I should learn some good basic science, and get my head out of the clouds. That Buck Rogers Stuff was fine for amusement—he read it himself—but in the real world . . .

In the real world I got a letter from that teacher, who had the honesty to send a note in August, 1969, apologizing to me and expressing gratitude that he'd not been able to discourage me from those crazy dreams. I wish he were alive so I could find out his reaction to *this* chapter.

It's not crazy dreams. It's not even Far Out. It's only basic engineering, and some economics, and a bit of hope. I may even have been too conservative. It probably won't take a hundred years.

Given the basic space civilization I've described, we'll have accomplished one goal: no single accident, no war, no one insane action will finish us off. We won't *have* to have outgrown our damn foolishness to insure survival of the race. Perhaps we'll all be adults, mature, satisfied with what we have, long past wars and conflicts and the like; but I doubt it. At least, though, there will be no way to exterminate mankind, even if we manage to make the Earth uninhabitable; and it's unlikely that any group, nation, or ideology can enslave everyone. That's Worth Something.

One suspects, too, that there will be an *enormous* diversity of cultures. Travel times between various city-states— asteroid, Martian, Lunar, O'Neill colony, Saturnian forward base, Jovian Trojan Point—will be weeks to months to years with presently foreseeable technology. That's likely to change, but by the time the faster travel systems are in widespread use the cultural diversities will be established. Meanwhile, communication among all the various parts of the solar system will be simple and relatively cheap, so that there will have been that unifying influence; cultures will become different because people want to be different, not because they don't know any better.

OK. In 100 years we'll have built a space civilization. We'll no longer have really grinding poverty, although there will undoubtedly be people who consider themselves poor, just as we have today people who live better than the aristocrats of 1776, but who think themselves in terrible straits. We'll have insured against any man-made disaster wiping out the race. What's next besides more of the same?

Why, we haven't even got started yet! "Be fruitful and multiply, and fill the face of the Earth," said the command; soon that will have been done; and some day we'll even run up against a filled solar system.

The first step is obvious. We can begin taking some of the more useless planets apart. They've got all that lovely mass, and it's concentrated so that we can't use it; better to make proper use of, say, Jupiter, and Mercury, and someday perhaps even Mars and Venus despite our having terraformed them.

At a thousand tons of mass per person, Mercury, taken

apart, could provide living space for 3×10^{20} people—that's 300 billion billion, rather a large population. People in the US at present dispose of about 10^{18} ergs per capita each year; small potatoes for a space civilization. Let's figure that our space people will need a million times that much, 10^{24} ergs per each per year, or a total of 3×10^{44} ergs for the people living on the skeleton of Mercury.

It's too much. The Sun only puts out 2×10^{39} ergs each year, and we can't catch all that. It seems we'll run out of energy before we run out of mass, and that's handy. Back to energy conservation! To support a really large population, though, we'll have to destroy some matter. Obviously that can't go on forever: so, while we're destroying matter, we may as well go elsewhere.

The stay-at-homes will busily take planets apart for their mass, and so fill space with flying cities that they'll soon catch great quantities of solar energy. You can just hear the asteroid civilizations (what's left of them) complaining about those closer in taking up all the light. Perhaps the Rockrats will be the first to say the Hell with it and leave, looking for a place to live where there's *elbow room*. Just too crowded in the solar system. "Not like when I was a kid, Martha. Not room to swing a cat nowadays."

They can take their whole civilization with them. The negotiations may take some time; the homebodies aren't going to want to let all that nice matter leave the system forever. Perhaps the Rockrats will promise to send back a nice fat planet from wherever they're going. It will take a while to pay off the debt, but they can pay it back with very high interest.

The trip will take many years, but so what? The Rockrats have taken their civilization with them. They'll miss the Sun, and by the time they arrive they'll have used up most of their asteroid, but by then people will live long lifetimes—and they'll darned well know how to exploit the new stellar system. "We'll do it right, Martha! None of those upstart places like Freedonia!"

Of course they'll already know about the planets in their new system. There's no real limit to the size of telescope you can build in space, and no problem about seeing; and with the lengthy baseline of the orbit of Ceres, or Jupiter's Trojans, or a Saturnian moon, astronomers will long since have discovered all the planets of all the nearby stars. There will probably have been probes sending back high-resolution pictures and making certain our colonists aren't heading for an already-occupied system.

And so it goes; across the Galaxy, as mankind fills system after system, and somebody begins to feel crowded. You'll note I haven't even postulated faster-than-light travel; I have given us matter annihilation, although that's not strictly necessary.

And beyond that? When we've tapped all the resources of easily available planets, and are still running out of metals and just plain mass? Well, there are stars—

Take an old star. A red giant, perhaps. Useless. No planets left—all consumed in the nova explosion that formed an ordinary star into a red giant. The poor thing is doomed in a few million years anyway; why not hurry it along? When it blows up, it will give off all kinds of useful materials.

Of course the star is a long way from civilization. The minerals could be picked up after the explosion, but maybe there's a better way: bring your planet-sized spacecraft reasonably near the target star. Turn on the matter annihilators and focus the resulting energy into a rather powerful laser beam. Shine it properly on the star. That's what you're going to do to blow it up anyway, but if you're selective enough about it you can turn the star itself into a rocket. Heat up this side, let it spew out starstuff, and it will move. Granted that's a slow process, and perhaps there'll be no economic incentive; but stranger things have happened in history. After all, the expedition will save its parent civilization; and life aboard the control planet need not be any more dull than, say, living in a colliery town; or going every day down to work at BBD&O . . .

But we needn't think about moving stars, or travelling to other stellar systems, anymore than Columbus and the Vikings had Cape Canaveral in mind. For the moment we need only concentrate on the next hundred years. There's quite enough to do right here.

In fact, I can just hear it now: "What good does it do to get people dreaming about that Buck Rogers Stuff? Why waste money on interstellar research when there's need for the money right here in the Trojan Points?"

Only One Earth indeed.

PART TWO:

STEPPING FARTHER OUT

COMMENTARY

One nice thing about writing science columns is that I can pick subjects that interest me. With all of science to play with, there's always something fascinating, and I never have to grind away on a dull topic simply because it's due.

All very well until it comes time to collect these essays into a book. Publishers insist that books have a central theme, or failing that, at least be neatly organized. Fortunately, when I began work on this collection, I found that most of the columns organized themselves into sections, each with a definite theme. Survival with style, black holes, space flight, energy; all relatively cohesive topics.

But there were a few leftovers. For the life of me I can't think of a theme to unite essays on holographic brains, flying saucers, terraforming Venus, and interstellar empires. It seemed a shame, though, to eliminate them just because they didn't fit into a neat package with the others. Besides, they illustrate just how rich and varied our future can be—and that is the point of the book.

Here Come the Brains

Robots are a favorite science fiction theme. Another is the great computer, much smarter than a man, which one way or another takes over the world. Machine intelligence fascinates us.

Comparatively fewer stories deal with enhanced intelligence, mostly because that's very hard work: how do you write about a character who is very much smarter than you are? One theme I've been working on for years involves implants: you take a small transceiver and put it into a human head (or elsewhere in the anatomy if you like), wiring up the output of the receiver into the auditory nerves.

Now you have someone who can communicate by a kind of telepathy; not only with other similarly equipped humans, but with really large computers. In theory, at least, every bit of information known to mankind will be instantly available to this "terminal man."

Dossiers; reference books; dictionaries; encyclopedias; company records; all the data banks of the government; IRS files; any of this can be his for the asking. A detective could get continuous information on the whereabouts of his colleagues, or on the personal habits of the suspect he's questioning. A company president has but to think the question to know about production, sales, and schedules.

There would be more: all the mathematical capability of powerful computers available in real time. Solve integral equations in your head, calculus of finite motion, orbits, stock market projections, all instantly available at a thought.

It is not very far-fetched; in fact, the concept as given above is too tame. There's no real reason to restrict ourselves to the comparatively inefficient input device of the auditory nerve. Why not simply crack the code used by the brain and squirt in the information? More on that later; for now, what I've described could probably be built today. Prosthetic surgeons *already* can wire a hearing aid directly into the auditory nerve.

The only problem would be the language used to communicate with the computer. Voice communication in ordinary English is not yet accomplished—although computers can be taught to recognize a surprisingly large vocabulary. My own computer doesn't have a voice board yet, but it would, I am told, be simple to attach one that would let the machine I'm writing this on recognize some 64 different words and commands.

In fact, it is nearly inevitable that before the end of this century there will be at least some humans equipped with the transplant transceiver I've described.

Unfortunately, it's *very* difficult to think like a man who has a 360/95 in his head. I don't recall too many memorable stories in which the real geniuses were the viewpoint characters, for the obvious reason that the author can't think as would the super-intelligent character. Of the two that impressed me the most, "Flowers for Algernon" (movie version was called *Charly*) and Ted Sturgeon's "Maturity," the central character lost the genius ability before the story ended.

And if there's a mental hookup to a computer in the future of some of the younger readers—and there probably is—there's another possibility also. A robot can be connected to the central brain, and of course there have been a *lot* of stories on that theme.

The only problem is, the central brain doesn't yet exist. Let's come back to that later.

* * *

There's no reason we couldn't build a central brain, but there's another approach: we may be on the way to *real* robots; self-contained, not relying on any kind of link with a central data bank, although able to use one if it's available; very strong; and capable of independent action if not thought. Again, the mechanics are not complex. The Artificial Intelligence lab at MIT has had some problems trying to build a robot arm as dextrous as a human arm and hand, but given enough money it could be done. What's missing is the brain.

The human brain weighs, on average, about 1.48 kilograms, or 3 1/4 pounds, in the mature male. (Chauvinists may amuse themselves with the fact that female brains run about 10% less in weight; but they're advised to do it privately.) That little chunk of matter can store some *one million billion* bits of information, which is quite a lot; the best computers don't yet have anything like that capacity, and they're still pretty big.

The computers are getting smaller all the time, though. I recall back in the early fifties visiting the ILIAC at the University of Illinois. ILIAC was at that time the biggest and most powerful computer in the world. Time on it was scheduled months in advance, and was reserved for the Naval Research Laboratories and the Institute for Advanced Studies, and such places. The computer was housed in a former gymnasium and cooled by the world's largest air-conditioner. Three undergraduates with shopping carts were employed full-time, three shifts a day, running around inside ILIAC's innards to replace burned-out vacuum tubes. Every computation was done three times and ILIAC took a majority vote on which was the correct answer—computations were slow, and tubes could burn out while they were going on.

It was an impressive sight. Nowadays, though, I can carry on my belt a TI 59—which is an order of magnitude more powerful than was ILIAC, and a very great deal more reliable.

Every year since the 1950's the information storable in a given calculator chip volume has *doubled;* and that trend shows no sign of slowing. So, although the human brain remains, despite all the micro-chip technology, the most efficient data-storage system ever built, electronics is catching up. And the brain is nowhere near as reliable as are the computers of today. Our brain does, though, have the capability for packing a lot of data into a small space and retrieving it quickly.

The brain has another characteristic that's very useful: the information doesn't seem to be stored in any specific place. Karl Lashley, after 30 years of work trying to find the engram—the exact site of any particular memory—gave up. All our memories seem to be stored all over our brains.

That is: Lashley, and now others, train specific reflexes and memory patterns into experimental animals, then extirpate portions of their brains. Take out a chunk here, or a chunk there: surely you'll get the place where the memory is stored if you keep trying, won't you?

No. Short of killing the animal, the memory remains, even when up to 90% of the cortical matter has been removed. Lashley once whimsically told a conference that he'd just demonstrated that learning isn't possible.

The experiment has been duplicated a number of times, and the evidence of human subjects who've had brain damage as a result of accidents confirms it: our various memories are stored, not in one specific place, but in a lot of places; literally, all over our cortices. That's got to be a clue to how the brain works.

A second characterisitc of the brain is that it's *fast.* Consider visual stimulation as an example. You see an unexpected object. You generally don't have to stop to think what it is: a hammer, a saucer, a pretty girl, the Top Sergeant, ice cream cone, saber-toothed tiger about to spring, or whatever; you just *know,* and know very quickly.

Yet the brain had to take the impulses from the light pattern on the retina and do something with them. What? Instrospection hints that a number of trial and error operations were conducted: "test" patterns were compared with the stimulus object, until there was a close correspondence, and then the "aha!" signal was sent. If, somehow, the "aha!" was sent up for the wrong test pattern, it takes conscious effort to get rid of that and "see" the stimulus as it should be seen.

We're still trying to teach computers to recognize a small number of very precisely drawn patterns, yet yesterday I met a man I hadn't seen for ten years and didn't know well then, and recognized him instantly. Dogs and cats do automatically what we sweat blood to teach computers. If only we could figure out how the brain does it ...

A number of neuro-scientists think they've found the proper approach at last. It's only a theory, and it may be all wrong, but there is now a lot of evidence that the human brain works like a hologram. Even if that isn't how *our* internal computer works, a holographic computer could, at least in theory, store information as compactly and retrieve it as rapidly as the human brain, and thus make possible the self-contained robots dear to science fiction.

The first time Dr. David Goodman proposed the holographic brain model to me, I thought he'd lost his mind. Holograms I understood: you take a laser beam and shine part of it onto a photographic plate, while letting the rest fall on an object and be reflected off the object onto the film. The result is a messy interference pattern on the film that, when illuminated with coherent light of the proper frequency, will reproduce an image of the object. Marvelous and all that, but there aren't any laser beams in our heads. It didn't make sense.

Well, of course it does make sense. There's no certainty that holography is the actual mechanism for memory storage in human beings, but we *can* show the mechanism the brain might use to do it that way. First, though, let's look at some of the characteristics of holograms.

They've been around a long time, to begin with, and they don't need lasers. Lasers are merely a rather convenient (if you're rich enough to afford them) source of very coherent light. If you don't have a laser, a monochromatic filter will do the job nicely, or you can use a slit, or both.

A coherent light beam differs from ordinary light in the same way that a platoon of soldiers marching in step differs from a mob running onto the field after the football game. The light is all the same frequency (marching in step) and going in the same direction (parallel rays). Using any source of coherent light to make a hologram of a single point gives you a familiar enough thing: a Fresnel lens, which looks like a mess of concentric circles. Holography was around as "lenseless photography" back before WW II.

As soon as you have several points, the neat appearance vanishes, of course. A hologram of something complicated, such as several chessmen or a group of toy soldiers, is just a smeared film with strange patterns on it.

Incidentally, you can buy holograms from Edmund Scientific or a number of other sources, and they're fascinating things. I've even seen one of a watch with a magnifying glass in front of it. Because the whole image, from many viewpoints, is stored in the hologram, you can move your head around until you see the watch *through* the image of the magnifying glass—and then you can read the time. Otherwise the watch numerals are too small to see.

Hmm. Our mental images have the property of viewpoint changes; we can recall them from a number of different angles.

Another interesting property of holograms is that any significant part of the photographic plate contains the *whole* picture. If you want to give a friend a copy of your hologram, simply snip it in half; then you've both got one. He can do the same thing, of course, and so can the guy he gave his to. Eventually, when it gets small enough, the images become fuzzy; acuity and detail have been lost, but the whole image is still there.

That sounds suspiciously like the results Lashley got with his brain experiments, and also like reports from soldiers with severe brain tissue losses: fuzzy memories, but all of them still there. (I'll come back to that point and deal with aphasias and the like in a moment.)

Holograms can also be used as *recognition filters*. Let us take a hologram of the word "Truth" for example, and view a page of print through it. Because the hologram is blurry, we can't read the text: BUT, if the word "truth" is on that page, whether it's standing alone or embedded in a longer word, you will see a very bright spot of light at the point where the word will be found when you remove the filter.

The printed word can be quite different from the one used to make the hologram, by the way. Different type fonts can be employed, and the letters can be different sizes. The spot of light won't be as bright or as sharp if the hologram was made from a type font different from the image examined, but it will still be there, because it's the *pattern* that's important.

The Post Office is working on mail-sorting through use of this technique. Computers can be taught to recognize patterns this way. The police find it interesting too: you can set up a gadget to watch the freeways and scream when it sees a 1964 Buick, but ignore everything else; or examine license plates for a particular number.

There's another possibility. Cataracts are caused by cloudy lenses. If you could just manage to make a hologram of the cataracted lens, you could, at least in theory, give the sufferer a pair of glasses that would compensate for his cataracts. That technique isn't in the very near future, but it looks promising.

You'll have noticed that this property of holograms sounds a bit like the brain's pattern-search when confronted with an unfamiliar object. A large number of test patterns can be examined "through" a hologram of the stimulus object, and one will stand out.

Brain physiologists have found another property of the brain that's similar to a holographic computer. The brain appears to perform a Fourier transform on data presented to it; and holograms can be transmitted through Fourier-transform messages.

A Fourier transform is a mathematical operation that takes a complex wave form, pattern, signals, or what have you, and breaks it down into a somewhat longer, but precisely structured, signal of simpler frequencies. If you have a very squiggly line, for example, it can be turned into a string of numbers and transmitted that way, then be reconstructed exactly. The brain appears to make this kind of transformation of data.

Once a message (or image, or memory) is in Fourier format, it's easy systematically to compare it with other messages, because it is patterned into a string of information; you have only to go through those whose first term is the same as your unknown, ignoring all the millions of others; and then find those with similar second terms, etc., until you've located either the proper matching stored item, or one very close to it. If our memories are stored either in Fourier format or in a manner easily converted to that, we've a mechanism for the remarkable ability we have to recognize objects so swiftly.

* * *

So. It would be convenient if the brain could manufacture holograms; but can it, and does it?

It *can:* that is, we can show a mechanism it could use to do it. Whether it does or not isn't known, but there don't appear to be any experiments that absolutely rule out the theory.

There are rythmic pulses in the brain that radiate from a small area: it's a bit like watching ripples from a stone thrown into a pond. Waves or ripples of neurons firing at precise frequencies spread through the cerebrum. These, of course, correspond to the "laser" or coherent light source of a hologram. Beat them against incoming impulses and you get an electrical/neuron-firing analog of a hologram.

Just as you can store thousands of holograms on a single photographic plate by using different frequencies of coherent light for each one, so could the brain store millions of billions of bits of information by using a number of different frequencies and sources of "coherent" neuron impulses.

That model also makes something else a bit less puzzling; selective loss of memory. Older people often retain very sharp memories for long-past events, while losing the ability to remember more recent things; perhaps they're losing the ability to come up with new coherent reference standards. Some amnesiacs recall nearly everything in great detail, yet can't remember specific blocks of their life: the loss or scrambling of certain "reference standards" would tend to cause *en bloc* memory losses without affecting other memories at all.

Aphasias are often caused by specific brain-structure damage. I have met a man who can write anything he likes, including all his early memories; but he can't talk. A brain injury caused him to "forget" how. It's terribly frustrating, of course. It's also hard to explain, but if the brain uses holographic codes for information storage, then the encoder/decoder must survive for that information to be recovered. A sufficiently selective injury might well destroy one decoder while leaving another intact.

In other words, the model fits a great deal of known data. Farther than that no one can go. The brain *could* use holograms.

Not very long ago, Ted Sturgeon, A. E. van Vogt, and I were invited to speak to the Los Angeles Cryonics Society. That's the outfit that arranges to have people quick-frozen and stored at the temperature of liquid nitrogen in the hopes that someday they can be revived in a time when technology is sufficiently advanced to be able to cure whatever it was that killed them to begin with.

I chose to give my talk on the holographic brain model. The implications weren't very encouraging for the Cryonics Society.

If the brain uses holographic computer methods, then the information storage is probably *dynamic*, not static; and even if a frozen man could be revived, since the electrical impulses would have been stopped, he'd have no memories, and thus no personality. If the holographic brain model is a true picture, it's goodbye to that particular form of immortality.

On the other hand, whether our own brains use holograms or not, holographic computers almost undoubtedly will work: and the holographic information storage technique offers us a way to construct those independent robots that figure so large in science fiction stories. Either way, it looks as if the big brains may be coming before the turn of the century.

* * *

The above was written in 1974. Surprisingly, it needed no revision, except to foreshadow what follows: Since 1974, there have been some exciting developments, most of which came to light at the 1976 meeting of the American Association for the Advancement of Science. They were reported in my column "Science and Man's Future: Prognosis Magnificent!", from which the following has been derived.

Studies of how we think—and of how machines might do so—continue. Take biofeedback. The results are uncanny, and they're just beginning. Barbara Brown, the Veteran's Administration Hospital physiologist whose book NEW MIND, NEW BODY began much of the current interest in biofeedback, is now convinced that there's nothing the eastern yogas can do that you can't teach yourself in weeks to months. Think about that for a moment: heart rate, breathing, relaxation, muscle tension, glandular responses—every one of them subject to your own will. Dr. Brown is convinced of it.

The results are pouring in, and not just from her VA hospital in Sepulveda, either. Ulcers cured, neuroses conquered, irrational fears and hatreds brought under conscious control—all without mysticism. When I put it to Dr. Brown that there was already far more objective evidence for the validity of the new psycho-physiological theories than there ever has been for Freudian psychoanalysis, she enthusiastically agreed.

One does want to be careful. There are many charlatans in the biofeedback business; some sell equipment, others claim to be "teachers." The field is just too new to have many standards, in either equipment or personnel, and the potential buyer should be wary. However: there is definite evidence, hard data, to indicate that you can, with patience (but far less than yoga demands) learn to control many allergies, indigestion, shyness, fear of crowds, stage fright, and muscular spasmodic pain; and that's got to be good news.

After I left the 1976 AAAS meeting in Boston I wandered the streets of New York between editorial appointments. On the streets and avenues around Times Square I found an amazing sight. (No, not *that;* after all, I live not far from Hollywood and thus am rather hard to shock.)

Every store window was filled with calculators. Not merely "four function" glorified arithmetic machines, but real calculators with scientific powers-of-ten notation, trig, logs, statistical functions, and the rest. Programmable calculators for under $300. *(Since 1976 the price of programmables has plummeted: you can get a good one with all scientific functions for $50 now; while the equivalent of my SR-50 now sells for $12.95 in discount houses. JEP)*

Presumably there's a market for the machines: which means that we may, in a few years, have a large population of people who really do use numbers in their everyday lives. That could have a profound impact on our society. Might we even hope for some rational decision-making?

John R. McCarthy of the Stanford University Artificial Intelligence Laboratories certainly hopes so. McCarthy is sometimes called "the western Marvin Minsky." He foresees home computer systems in the next decade. OK, that's not surprising; they're available now. *(Since that was written, the home computer market has boomed beyond anyone's prediction; in less than two years home computers have become well-nigh ubiquitous, and everyone knows someone who has one or is getting one. I even have one; I'm writing this on it. JEP)* McCarthy envisions something a great deal more significant, though: information utilities.

There is no technological reason why every reader could not, right now, have access to all the computing power he or she needs. Not wants—what's needed is more than what's wanted, simply because most people don't realize just what these gadgets can do. Start with the simple things like financial records, with the machine reminding you of bills to be paid and asking if you want to pay them—then doing it if so instructed. At the end of the year it flawlessly and painlessly computes your income tax for you.

Well, so what? We can live without all that, and we might worry a bit about privacy if we didn't have physical control over the data records and such. Science fiction stories have for years assumed computer controlled houses, with temperatures, cooking, menus, grocery orders, etc., all taken care of by electronics; but we can live without it.

Still, it would be convenient. *(More than I knew when I wrote that; I don't see how I could get along without my computer, which does much of that, now that I'm used to it. JEP)*

But what of publishing? McCarthy sees the end of the publishing business as we know it. If you want to publish a book, you type it into the computer terminal in your home; edit the text to suit yourself; and for a small fee put the resulting book into the central information utility data banks.

(So far I have described how I now, only two years after I wrote the above, prepare my own books. The difference is that after I have them composed on the TV-like screen, and edited to my satisfaction—a computer-controlled type-writer puts it onto paper, which is mailed to New York, edited again, and given to someone to type into elec-tronically readable form for typesetting. Obviously that stage will be eliminated soon; why can I not send a tape and be done with it? Incidentally, the NY Trib had no type-writers or paper at all: reporters and rewrite persons worked on a TV screen, editors called that up to their screens, and when done the text went directly to compos-ing without ever being on paper at all. JEP)

Once a book is in the central utility data banks, those who want to read it can call it up to their TV screen; a royalty goes from their bank account to the author's; where is the need for printer or publisher? Of course some will still want *books* that you can feel and carry around; but a great deal of publishing can be as described above, and for that matter there's no reason why your home terminal can-not make at reasonable cost a hard copy of anything you really want to keep.

Few publishers own printing plants; most hire that done. What publishers provide is editorial services and dis-tribution. The latter function will largely vanish: the information utility does that job. There remain editorial services.

With such a plethora of books as might appear given the above—after all, the only cost to "publish" a book would be to have it typed, plus a rather nominal fee to the utility for storing it—critics and editors will probably grow in importance. "Recommended and edited by Jim Baen," or "A Frederick Pohl Selection" would take on new significance, and one assumes that these editors would continue to work with authors since they'd hardly recommend a book they didn't like (and some authors might even admit that a good editor can help a book). "Big Name" authors would probably have little to worry about, with their readers setting in standing orders for their works; new writers would probably have to get a "name critic" to review their stuff.

OK; still not all that new for veteran science fiction readers; but did you catch the time scale? The equipment, *all* of it, exists *now*. The telephone net to link nearly everyone in the US with the information utilities exists *now*. Computer electronics costs are plummeting. McCarthy's home terminal can be with us in the next five years, with the information utility fully developed in ten to fifteen.

In fact, the only obstacle is entrepreneurial: the equipment and technology exist at affordable costs. It takes only someone to organize it.

But—in twenty years we may not need the home terminals except as backup. Dr. Adam Reed of Rockefeller University has a new scheme: direct computer to brain hookups.

Ten years: Dr. Reed believes that within ten years we will have cracked the code that the brain uses for information processing and storage. Once that's done, information can be fed directly into the brain's central processing unit without going through such comparatively slow peripheral equipment as eyes and ears. You need not read a book: the computer can squirt the book's contents directly into your mind.

Of course it won't be the same experience: that is, when I read *War and Peace* there was more than a transfer of information. There were also emotional responses. Those would be lacking in the direct information-acquisition experience. Thus there will probably remain a few nuts who read, just as TV hasn't quite eliminated literacy in the US; but it may well be that within your lifetime the normal method of acquiring information, particularly of grasping the content of dull books that everyone wants to have read but no one wants to read, will be through computers.

This means a complete restructuring of our education system, and perhaps it is high time; yet I have met few teachers who have thought about the new capability at all, and there is no one I know of planning for the time when we do not have to sit in classrooms for the first twenty years of our lives.

There will always be a need for education, of course; for those who can teach their pupils to *use* the information available to them; and who will teach them to be civilized (although that latter may not be a function of schools, and certainly is only indifferently performed in large areas just now.)

Incidentally: Reed believes that each of us has a different code; not all brains use the same information processing symbols. Thus each of us would need a computer that has been taught to use *our* coding system. That is no bar, of course; the computer system need not be very expensive, and probably won't be, at least not after a few years. (One speculation: if each of us uses a different coding system, then true telepathy would be rare—and far more common among identical twins than among others. All of which seems to echo experience.)

And the implications of all this are staggering. In the near future—in *your lifetimes,* most of you—there will be those who, having obtained an implant, will quite literally know everything known to the human race. (This assumes that the information utilities will also exist; but those seem inevitable.) Want a multiple-regression equation linking weather, gasoline consumption, electricity generation, ship keels laid, the price of wheat futures, and the number of wall posters in Peking? Merely think the question and wait; it shouldn't be long before you have it.

Because, according to Reed, the implanted transceivers I have used in various stories (*High Justice, Exiles to Glory,* etc.) are perfectly workable—but, as mentioned earlier, I may have been too conservative. Certainly though we will have implants that "talk" to you, feeding information directly to auditory and optic nerves; in fact we have them, crudely, now, and use them to make the blind see and the deaf hear. So far have we come in the past few years. In the not distant future we shall do more for the handicapped than was ever thought possible. The "Bionic Man," shorn of some of the more impossible touches that violate the laws of thermodynamics, may become reality in this century.

But go further: when the coding system is completely known, a human personality can be "recorded"; and if the cloning experiments prove out, the personality can be transcribed into a younger edition of the same person: know what you have learned at fifty, or eighty, and put that into a body aged 25.

Far out? Science fiction? No. There's a very real possibility that it can happen to some of you; a very small possibility that it might happen to me.

It's getting hard for science fiction writers to keep up: even we are getting future shock. But it's all for real, you know.

It can all happen. The Big Brains are coming.

The Big Rain

Mankind needs frontiers. We need new worlds to conquer, impossible odds to overcome, a place of escape from bureaucracy and government; a place where life is hard but the problems are simple, requiring no more than courage, determination, and hard work to win great rewards.

Even for those who will never go chasing out to the frontier there's a great comfort in knowing it's *there:* that you could, if you chose, pull up stakes and try your hand at making a new life. For the warriors and dreamers among us a frontier is so vital that if there isn't a physical one, we'll create an internal problem and fight that.

I suppose that man's need for a frontier is a debatable proposition, in that somebody might question it; but I doubt that many science fiction readers would dispute the point. To a very great extent this is what science fiction is all about.

Unfortunately, science hasn't been cooperating with us. First comes Special and General Relativity to say that we won't ever travel to the stars. Then come the space probes to rob us of our traditional solar system. What's left?

As I've said before, I firmly believe we'll overcome the speed-of-light barrier, and if we don't, we'll still find a way to leave the Solar System; but that may take quite a while, and suppose I'm wrong. Are there no frontiers left?

Venus was once a favorite. Hot and swampy, a younger sister of Earth, with grey skies laden with thick clouds; primordial ooze, scattered thick forests burdened with heavy vines and hanging mosses; thick fungus that ate men alive; a world populated with strange animals, dragons and dinosaurs and swamp creatures resembling the beastie from the Black Lagoon, Venus was a world to challenge us.

Then the scientists took away our Venus, as they had taken away our Mars. For a time the extremely high temperatures of Venus gave some comfort to Velikovsky and his supporters, and thus argued for a more unstable and less orderly Solar System than we had imagined—we could take comfort in fright. In a world of cosmic catastrophes there is room aplenty for adventure and derring-do.

After all, Dr. V. had predicted that Venus would be very hot, possibly even still molten from her fiery birth from Jupiter; but, alas, even that is denied us. Pioneer looked down on Jupiter and found that he is not even solid. The Queen of Heaven does not resemble her mythological father, no, not in the least.

Velikovsky would have hydrocarbons (and carbohydrates) in the Venerian atmosphere, and he may well be right; but mostly there is carbon dioxide (CO_2) in fearful amounts, diluted with sulfuric and hydrofluoric acids. Here and there may lurk clouds of ice crystals and water vapor, but mostly there is a poisonous and corrosive brew pressing down with 90 atmospheres on Aphrodite's face.

Venus does not seem an attractive place.

* * *

When I was a boy I read a "juvenile" novel so utterly forgettable that I recall neither plot nor title nor author nor characters. One incident in that book impressed itself in my mind.

The heroes had stranded themselves in the Arabian (or Moroccan or Tunisian or Saharan) desert, and were about to be engulfed by wild barbaric tribesmen on camels—when lo!, the tribesmen retreated in panic, driven away by the sight of a regiment of British Tommies in full uniform.

There were, of course, no British troops for hundreds of miles. The author made a point of explaining that this sort of mirage happens quite often in the desert. I remember looking for one many years later; it's true enough.

There are numerous stories of ghost cities in the Mediterranean. One can often be seen across the straits of Messina: a full city, with moving traffic, nestled onto the dry and barren hills. The illusion has been known for at least three thousand years.

What happens is that changing air densities will affect light rays so that under the proper conditions the image is refracted over the horizon. A British regiment marching in Aden is seen a hundred miles away. Ghostly images of cities form on barren shorelines.

This is rare on Earth. On Venus it's inevitable. The Venerian atmosphere is so thick that light is refracted through 90° and more. The whole planet is visible from any point on the surface. The explorer will seem to stand in the bottom of an enormous bowl, with cliffs towering high above him.

This doesn't have a lot to do with the subject of this column, but it fascinates me. Venus must be a confusing place, where one sees the back of one's own head spread about on the top of an enormous cliff...

* * *

The mirages might be worth seeing, but there's not a lot else to attract colonists or tourists. Carl Sagan, Cornell U's resident genius and expert on Venus, once said "Venus is very much like Hell," and a glance at Figure 10 shows why.

It's just not a very attractive place to live. In fact, a more useless planet is hard to imagine. What good is a lump of desert whose surface temperature is up there about the melting point of lead, whose atmosphere is too thick, with winds of fearful velocity and force blowing dust across craters and jumbled structures like the surface of the Moon?

Some writers have proposed that since we can never visit Venus (and wouldn't want to if we could), we should make her the Solar System's garbage heap. Venus could become the repository for all the long-term radioactive wastes produced by nuclear power plants. It's cheaper to drop a load onto the surface of Venus than to send it into the Sun, and what other use do we have for Hell anyway?

Quite a lot: Venus is very likely to become the first terraformed planet. In a few hundred years there may be more people living on Venus than live on Earth at present.

The asteroids can be one frontier for the future, as I have described elsewhere; but they'll never be developed into a New World. That's reserved for Venus.

Not only can we terraform Venus, but we could probably get the job done in this century, using present-day technology. The whole cost is unlikely to be greater than a medium-sized war, and the pay-off is enormous: a whole New World, a frontier to absorb adventurers and the discontented. Few wars of conquest ever yielded a fraction of that.

Hardened veteran SF readers will have recognized the title of this chapter as coming from a 1955 ASTOUNDING novelette by Poul Anderson. His "The Big Rain" should rank with all the other successful predictions by SF writers: when the Big Rain comes, we can live on Venus.

The Venerian atmosphere consists mainly of carbon dioxide plus some other junk that we'd like to get rid of.

Figure 10

THE PHYSICAL PARAMETERS
OF HELL

	EARTH	VENUS
Diameter	7927 miles	7700 miles
	12,756 km.	12,392 km.
Mass	5.98×10^{27} gm (1)	4.9×10^{27} gm (0.82)
Density	5.52	5.27
Distance from Sun	1.49×10^{13} cm =	1.08×10^{13} cm =
	1 AU	0.7 AU
Length of Year	365.25 days	225 Earth days
Rotation Rate	23 hrs 56 min	243 Earth days
		(Retrograde)
Mean Solar Day	24 hours	118 Earth days
Mean Irradiance	1.97 cal/cm²/min	3.78 cal/cm²/min
Atmospheric Pressure	14.7 lbs./in² =	90 atmospheres
	1 atmosphere	
Mean Equatorial Temperature Range	0 to 50° C	- 20° C (cloud tops) to 480°C (Surface)
Surface Gravity	980.7 cm/sec² =	894 cm/sec² =
	1 g	0.9 g
Distance to Horizon	20 km.	38,000 km.

The junk is water soluble, and will wash away when we get the rain to fall.

That thick CO_2 blanket is the cause of all Venus's problems. Solar radiation comes in, a lot of it as visible light and hotter, into the ultra-violet. It penetrates into the atmosphere and is absorbed. As Venus slowly rotates, until she faces the absolute black of outer space, the absorbed radiation tries to go back out. But it has cooled off somewhat, from ultra-violet to infra-red. IR is absorbed nicely by carbon dioxide. The heat never gets back out, and up goes the Venerian temperature. This is called "the greenhouse effect" and works quite nicely for farmers on Earth, as well as out there.

In fact, there are theorists who wonder if burning all those fossil fuels won't loose so much CO_2 into our own atmosphere in a couple of generations as to bring Earth's temperature up sharply. The result would be melting ice at the poles, and the drowning of most of our sea-coast cities.

Before we get too alarmed at that, though, there's something else to worry about: it seems that far from rising, the average temperature of the Earth is *falling*, and we may be due for a new ice age, complete with glaciers in North America and like that, within the next hundred years. For more on both subjects see other chapters; just now we're concerned with reducing the Venerian fever to manageable levels.

We need to break up that thick CO_2 layer around Venus. This will automatically liberate oxygen. It will also chop down the atmospheric pressure to something tolerable.

Breaking up CO_2 is a rather simple task. Plants do it all the time. We'll need a pretty rugged plant, since the atmospheric temperatures of Venus range from below freezing to live steam, but fortunately one of the most efficient CO_2 converters is also one of the most rugged.

In fact, it's generally thought that the blue-green algae were responsible for Earth's keeping her cool and not getting covered with a thick CO_2 blanket like her sister.

The algae will need sunlight, water, and CO_2. There's no question about finding those on Venus. The temperatures are all right, too, at least in the upper atmosphere where we'll seed the algae.

This isn't just theory. "Venus jar" tests have shown that blue-green algae thrive in the best reproduction of the upper Venerian atmosphere we can make, breaking down the CO_2 and giving off oxygen at ever-increasing rates. In fact, to these algae, Venerian conditions are not Hell but Heaven.

We have the algae, and we can build rockets to send them. About a hundred rockets should do the job. Say each rocket costs 100 million dollars, and the ten billion dollar price doesn't seem unreasonable. Say we're off by a factor of ten, and we've a hundred billion, less than wars cost; and out of *that* much money we should be able to get a couple of manned Venus-orbit laboratories as a bonus. It will be, after all, a once-in-evolutionary-lifetimes opportunity.

Once the algae are sprayed into the upper atmosphere, they happily eat up CO_2 and give off oxygen. They have no competition. Nothing eats them, and there's plenty of room for expansion, plenty to eat, and lots of sunlight.

Some calculations show that within a year of the initial infection the surface of Venus will be visible to Earth telescopes. Meanwhile the algae go on doing their thing. The atmosphere clears. Sunlight coming in begins to radiate back out. The atmospheric temperature falls, and lower levels are invaded by the algae.

There is about 100 inches of precipitable water in the Venerian atmosphere. This means that if it all fell as rain, it would cover the entire surface to a depth of 100 inches.

This sounds like a lot until you contemplate the miles of water standing over most of the Earth. Still, 100 inches is respectable. Compare it to Mars, with 10 *microns* of precipitable water, and you'll see what I mean.

As the air above Venus cools, raindrops form. Eventually it will rain, and rain, and rain. The first planet-wide storm probably won't ever get to the surface: the rain will evaporate long before that. But as it evaporates, it cools still another layer of atmosphere; down move the algae.

Repeat as needed. In no more than twenty years from Go, the Big Rain will strike the ground. Craters will become lakes. Depressions will become shallow seas. Rivers will begin carving channels.

As Venus slowly turns, there may be snow on the night side. A water-table will develop. Deserts that may be a billion years old will turn to mud.

By now the surface will be tolerable to humans with protective equipment, and the seeding can begin. Scientists will want to move very carefully, introducing only the *right* plants and insects, fearful lest an unbalanced ecology result. Against this there will be pressure from colonists who want the job over and done with.

Some will demand that we dump a little of everything we can think of onto Venus and let competition take its course; an ecology will result inevitably, although we may not be able to predict what it will be.

There will also be tailored organisms. Microbiologists are already to the stage of switching genes from one species to another, and it shouldn't be long before this is done with higher plants and animals.

After all, Venus will have special conditions. That long rotation period means severe winters, like the Arctic tundra or worse. Much of Venus may resemble Siberia or the North Slopes of Alaska, which, if you haven't seen them, are second only to Antartica as candidates for the most desolate spots on Earth. On the other hand, Venus will still have a thick atmosphere, and she's closer to the Sun. We don't know what the final temperature will be, or how much heat-pumping the atmosphere can do.

It may not be the most pleasant world imaginable. Some writers have speculated that Venerian colonists will be nomads, staying on the move to live in perpetual sunshine. Others have described a world of paired cities connected by rails: as sunset approaches, the inhabitants escape the winter night by travel to their city's twin at the antipodes, somewhat as the Martian colonists migrated yearly in RED PLANET.

Much of this scheme was described by Carl Sagan in a 1961 article in *Science*. Poul Anderson amplified it in "To Build A World," *Galaxy*, 1964 (June). In Poul's story there was fierce competition for the better parts of Venus, resulting in a clan structure social system and innumerable limited wars between clans. It's a reasonable projection: the first settlements will be small, probably dominated by one man, and intermarriage may well result in clans.

We haven't yet mapped the face of Venus, so we don't know where or how large the seas will be. We don't know a lot (really, almost nothing) about the sub-structure of Venerian soil. How much water will be absorbed? At what level will a water-table form?

For that matter, will it be enough to introduce our algae to get everything started, or will we also have to provide fertilizers: phosphorus, trace elements, that sort of thing?

Our massive tampering with atmospheric energies and surface temperatures may trigger techtonic activities. Venus may erupt in a number of places, and spew out even more CO_2, water vapor, methane, and such like.

Project Morning Star won't be all smooth sailing. There will undoubtedly be unforeseen problems. For all that, the terraforming of Venus is no pipe dream. We could do it. We can do it right now, if we want to pay for it. We can create a new frontier, larger than ever was the New World.

In other words, there's room here for more stories than we had about the "old" Venus with her swamps and dinosaurs. True, we won't have any intelligent Venerians to contend with. It's unlikely that there's any life on Venus at all.

Unlikely but not impossible. There are, after all, Earth-like conditions of temperature and pressure in the Venerian atmosphere, and this is about the only planet—other than Earth—in the Solar System that can make that statement. Moreover, it may be that there have *always* been spots with Terrestrial conditions on Venus.

True, life isn't likely to have evolved under present Venerian conditions, but some planetary scientists now believe that Venus was once much more like Earth. Then, for reasons not completely understood, the planet began to heat up and dry out.

If by that time life had evolved, it may have taken to the air, and be hanging around there yet.

It isn't likely, of course. If Venus had really active bugs they should be busily tearing off the CO_2 cover that keeps Venus hot, and we wouldn't need to infect the Queen of Heaven with blue-green algae.

If there are any native Venerian bugs, Project Morning Star will probably doom them to extinction.

We have the technology to make Venus a place where we can live. Have we the right to do it? *Should* we, granted that we can?

After all, this is "pollution" on a grand scale. True it's not pollution to our way of thinking; but what's good clean air to us is certainly un-natural to Venus.

I suppose that we can and probably will debate this question for a long time to come, and when the debate is finished I suspect that no one will have changed opinion by one jot. Those who feel it monstrous to go about interfering with "nature," and see the terraforming of Venus as blasphemy (they will probably use the term "obscene") are not likely to change their minds.

Those who see Project Morning Star as the most glorious opportunity that has yet faced man are unlikely to be concerned about Venerian gas-bags or other hypothetical Venerian critters (beyond building them a zoo to live in).

Meanwhile, one thing is certain: it is possible that within my lifetime I could dateline this column "Venusberg." We can do it, and some of you could live there.

Just after the Big Rain.

In the years since the above was written we have gained considerably more knowledge of Venus. Conditions there are even more frightful than we thought, and the terraforming of Venus appears now to be more difficult than I knew.

It is still not impossible—and our capabilities have increased as well.

We could bring about the Big Rain if we wanted to.

Flying Saucers

Science fiction people, fan and professional alike, tend to avoid the subject of "flying saucers." After all, we're much more scientific than *that!*

I recall the first SF club meeting I ever attended. It was in Seattle, and the group was called The Nameless Ones. (They had a nasty habit of electing newcomers President at their first meeting; but that's another story.) For some reason a reporter showed up, and the first question she asked was about "flying saucers." The Nameless rather gruffly told her we weren't interested and never would be.

The reaction was probably justified. After all, we were those nutty people who wanted to go to the Moon, and in the 50's that was far out enough. How could we claim space travel was respectable if we were also saddled with flying saucers?

SF people have always tended to shy away from UFO's, and I've been no exception; but now the staid and stolid

American Institute of Aeronautics and Astronautics includes panels on UFO's in their annual Aerospace Sciences meetings. If the inheritors of the professional American Rocket Society can discuss UFO's in the normal language of dull science, maybe we SF dreamers ought at least to think about them.

What really got me onto the subject, though, was the publisher of the newspaper chain I write science columns for. He wanted a feature on UFO's. I took the assignment with a certain degree of fear.

In my previous experiences, UFO enthusiasts were invariably wild-eyed, generally insisted that I look at smudgy photos, and often revealed that the US Air Force was engaged in a conspiracy to suppress all knowledge about UFO's. They told horror stories about Project Blue Book. They solemnly related that the US Government had constructed a secret laboratory in the Mojave, and seduced a famous UFO investigator into thinking it was an extra-terrestrial space ship so that later they could embarrass him.

I had been told of hundreds of excellent photographs seized by the USAF Blue Book officers, taken away and never to be returned despite vigorous legal efforts to recover them—but somehow had never been given the name of the lawyer who filed the suit, the court in which it had been filed, or the judge who heard it.

I also remembered a couple of USAF captains I'd worked closely with when I was in the space program, and their stories about Blue Book. Blue Book was a "George" job (there's nobody to do it? Give it to George.) which everyone started off conscientiously and soon began to hate as the silly and inconsistent stories poured in.

However, an assignment is an assignment, and I dutifully looked up and interviewed as many UFO experts as I could find.

The field turns out to be more interesting, and far more respectable, than I would have thought.

* * *

Interest in and study of Unidentified Flying Objects—UFO's—is no longer confined to fanatics and eccentrics, if indeed it ever was. I don't mean to imply contempt for *all* the early investigators, or for the amateur outfits like MUFON and NICAP who collect the bulk of the data on the subject. However, the professional scientists have also moved into the field.

The Dean of UFO scientists is Dr. J. Allen Hynek, Chairman of the Department of Astronomy at Northwestern, and Director of the Center for UFO Studies (Box 11, Northfield, Illinois, 60093).

The consultant list for the Center includes such notables as Dr. Claude Poher, one of the Directors of the French equivalent of NASA, at least one Nobel Laureate, and any number of random PhD's in various sciences. Dr. Hynek himself looks like a very conservative astronomer, which in fact he is. He was originally hired by USAF as a UFO consultant, and began with the opinion that UFO reports were nonsense to be explained away. Unlike some others, notably the late Dr. Edmund Condon of Colorado U., Hynek didn't keep that view.

He now hopes that some progress on UFO research may be made during his lifetime, and views his Center as his scientific legacy. His book , THE UFO EXPERIENCE , is still the best general work on the subject. One sign of the increasing respectability of UFO studies is that Hynek's book was favorably reviewed by planetologist Bruce Murray in the AAAS journal, *Science*. (Not that Murray is a UFO enthusiast; far from it; but he took the trouble to examine the subject before writing a review. Alas, such courtesy seems even more rare in the scientific professions than among writers.)

Hynek's Center is now tied in with many law enforcement agencies and maintains a toll-free number available to police. Officers across the country can report UFO sightings and get advice on disposition of UFO cases.

This came about largely because the FBI published a

long article on UFO's in the February 1975 issue of the FBI *Bulletin*. As Hynek says, that's practically "the Good Housekeeping Seal of Approval."

All right. UFO research is respectable, if not orthodox. Now what the devil are UFO's?

No one knows. There are plenty of speculations, but very little evidence. We'll go into some speculations in a moment, but first let's see what we're discussing.

First, we can say what UFO's probably are not: namely, they are not misinterpretations of "usual" or "ordinary" phenomena. There are plenty of such, of course, but by definition if we can *identify* the cause, we don't have a UFO.

Incidentally, this is the major failure of the Air Force financed Condon Report. Condon never investigated a single case, and chose to concentrate nearly all his efforts on known mistakes and misidentifications.

In fact, Condon even sought out people like the "man from Galaxy Three" who wanted $100,000 "to build runways on orders from Galaxy Control." Condon's administrator put out a memo stating that the purpose of the study was to explain away UFO's, but to make it appear that a scientific investigation had been carried out. There was great concern that the staff would be laughed at by orthodox scientists, and efforts made to show that no one in the study really took it seriously.

Thus the Condon study never did do what the taxpayers put up their money for, namely, investigate *unidentified* flying objects. It does a pretty good job of showing the kinds of mistakes that have been made, but as a scientific study it is valueless. On the other hand, it probably served its major purpose, to get UFO's out of the Air Force's hair. (USAF had for years tried to give UFO studies to *someone:* NSF, the Weather Bureau, Air Defense Command [Army], National Academy of Sciences, anyone who'd take it, budget and all.)

Yet when we've got rid of the kooks and cranks, mercenaries and swamp gas and meteorological balloons, the planet Venus, helicopters, and hoaxes, there remain cases that we cannot explain. Hundreds of them, with nearly a hundred reported by multiple witnesses of presumed good honesty and integrity.

The USAF Assistant Chief of Staff, Intelligence, said back in 1947 that "credible observers are reporting incredible things." Nearly 30 years later that's a good summary. The observers are credible by any test. The reports defy belief.

I want to emphasize something: we have convicted people of murder on far less evidence than we have for the existence of "incredible" UFO's. Our legal system routinely tries to sort out fact from fancy, and to examine such intangibles as "honesty and integrity." However well it works or doesn't work, courts regularly try cases on flimsier evidence than we have in the UFO reports, and hear witnesses far less credible than those Hynek has singled out for his studies.

If you were on a jury you'd be likely to believe the people Hynek has interviewed. He excludes almost all of the famous UFO "investigators" who grow wealthy from their UFO tales.

Hynek's classification scheme is as good as any. He sorts UFO reports into the following categories: Daylight Discs, Nocturnal Lights, Radar-Visual, and Close Encounters of the First, Second, and Third Kinds.

The first two are simple enough. They also exhibit a number of similarities: rapid to enormous velocities and accelerations, no sonic boom despite high velocity, etc. Radar-Visuals are those reported by both kinds of observation on the same phenomenon, usually by highly professional personnel such as USAF radarmen, professional air traffic controllers, etc.

So far so good. Were these three the whole of it, we could comfort ourselves with the thought that there's probably an explanation well within the limits of present-day science.

Unfortunately they are not the whole of it.

The Close Encounters are disturbing, but there's a lot of reliable evidence for them: reliable, that is, in that the observers would be believed if they told nearly any other story. Close Encounters of the First Kind involve observations at ranges of 20 to 500 feet, close enough to see details.

Close Encounters of the Second Kind involve some physical effect on the observers or their surroundings: interference with auto ignition (a common report); movement of trees, as was photographed in the famous Oregon disc; or, sometimes, thermal and physiological effects.

Close Encounters of the Third Kind involve inhabitants, generally humanoid. If we have trouble swallowing the first two encounters, this one really makes us want to gag; yet, again, the reports are about as good as those given in criminal courts, or sent out by war correspondents, or indeed, for most of us, for the existence of any other complex phenomena we haven't ourselves seen.

Beyond this point Hynek and most UFO scientists draw the line. There are reports of actual communication with UFO's, many given by people who *seem* to be telling the truth, but first there are few such, and secondly, those making them nearly always manifest some kind of psychological aberration. We can note that the experience itself might be enough to unhinge most of us, and still confine our work to the three kinds of close encounters, the discs, the night lights, and the radar visuals.

* * *

OK. That's the subject matter. Reports, sometimes accompanied by photographs, sometimes not. There aren't a lot of photographs, and of those not many have been or can be checked and pronounced unmistakably genuine; *but there are some*. There are others which *may* be genuine, but can't be proved to be; but those unmistakably genuine are disturbing enough.

Now what do we mean by genuine? Well, among other things, that the negative exists, so that photo experts can be certain this isn't either a double exposure or some kind of fakery from the printing lab; that there's a real object recorded on the film.

Next, they want to see other objects besides the UFO: trees, houses, wheat fields, etc., so that the distance to the UFO, and thus its size, can be established. This generally takes care of thrown objects and the like. The experts are even happier with a series of photographs, because they can take the sun angle off each one, and again eliminate a lot of thrown or suspended objects.

I won't go into all the tests because I'm not a photo expert. I did conduct a long interview with Adrian Vance of *Popular Photography* magazine, and also with some USAF professionals, and I'm now convinced: there exist several photographs of genuine objects, taken at distances of some 50 to 500 feet. The objects are in flight. They tend to be circular, and of dimension about 30 feet diameter by 7 feet at the thickest point. At least one (the Oregon "saucer") had a photographable effect on very large trees.

I find it hard to believe that's a thrown or suspended object. Moreover, in several unrelated cases of photographed discs there were multiple observers with no obvious connections with each other and no discoverable reason for making up the story. (As is usual in most cases, the observers do *not* want their names in the paper, do *not* want to be paid for their information, and are *not* interested in going on lecture tours.)

OK again. Some readers have always "believed" in UFO's. Some others may now be convinced there's *something* there, and rather a larger number are probably convinced that *I* believe there's something there.

So what are they?

Gee, I don't know. I used to say I was uninterested in UFO's because they just couldn't be intelligent critters. Why couldn't they? Because there was no place they could have come from. Earth? Not really likely. The Solar System? Unlikely a few years ago, virtually impossible now given what we've learned about the other planets.

Another star system, then. Now that *really* raises problems. How do they get here? Faster than light travel? Science fiction aside, although the General Theory of Relativity isn't anywhere near universally accepted, the Special Theory forbids faster than light travel by material objects, and gets more and more corroboration every year.

But then, so what? Do I really "believe in" the absolute limit of the speed of light? No. I accept it as a probable working hypothesis, but I firmly hope to see faster than light (FTL) travel in my lifetime. I *hope* to see it; but I can't tell you how it will happen, and the evidence is all against me. Still, I do not rule out FTL as impossible, and thus I can't say interstellar visitors are impossible either.

This is the point at which scientists get nervous. Not only Hynek, but men like Dr. Robert Wood (BSEE, Aeronautical Engineering; PhD, Physics, Cornell) who is an engineering manager for a large aerospace firm he'd rather not have named in an article on UFO's; a Nobel Laureate who'd rather not be named at all; all of them say almost exactly the same thing when you ask, "What do you think UFO's are?"

They say: "You want some wild guesses? Hypotheses we'd be willing to defend at a scientific meeting? Science fiction? Where's this going to be published, anyway?"

Nervous indeed. Every one of them wants it clearly understood they're talking hypotheses, theories which not only may not be true, but probably aren't true; speculations, if you will.

Begin with Dr. Wood. "I'd now be willing to defend the extra-terrestrial hypothesis at an AAAS or AIAA panel meeting."

Continue with Dr. Hynek: "They seem to act intelligently. I wouldn't be surprised to see proof that they're ob-

121

servers from another stellar system. I don't say that's what they are, or that I 'believe' in little men from outer space; but I wouldn't be shocked to see it proved."

Continue further with Hynek. We ask how they got here, given special relativity and all that. "People want answers, but we have none to give. Yet I suspect we may be witnessing something as far beyond us as, say, television would have been to Plato. After all, not too many years ago the best minds in the world couldn't explain the Northern Lights. They were there. You could observe them. But we hadn't the basic science to begin to understand them. I think that's what we've got here."

In fact, it is precisely this that interests Dr. Hynek, and probably most other UFO scientists as well. Understanding UFO's may lead us to new sciences far beyond anything we have or can even imagine. It is this which leads Hynek to hope the UFO Center he founded will be his scientific legacy.

What are some ways the UFO's and their hypothetical inhabitants might achieve FTL? Hynek: "Perhaps they aren't material in the way we think; that they use something like television to transport themselves. Matter transportation, whether instantaneous or at very high velocity."

Robert Wood: "I haven't a clue. Yet there's been a small revival of interest in ether theories lately, have you noticed? Perhaps Special Relativity doesn't hold after all.

"Also, there's the possibility of gravitational interactions. Or try this: these things are generally reported accompanied by a really overwhelming magnetic field. What IS the speed of light in a billion gauss magnetic field? Whatever it is, I'm sure there's no magic here, and it may be consistent with science we are about to discover ourselves."

Hmm. These are scientists, not SF writers.

In other words, although we "know" that UFO's with intelligent inhabitants can't come from our solar system

122

and can't get here from any other, we've "known" a few other things that turned out not to be so. It hasn't been all that long since we "knew" the Law of Conservation of Matter; that atoms couldn't be split; that the Sun couldn't possibly have been burning for longer than a few thousand, uh, hundred thousand, well, we can *prove* not longer than a million years; that heavier than air craft couldn't fly, certainly couldn't fly faster than the speed of sound; that nothing could get into outer space from here, well, not very *far* out, anyway; and so forth. Now some things we "know" turn out to be true, and maybe Special Relativity will be one of them; but maybe it won't, either.

Assume the UFO's employ a technology capable of interstellar travel. What the devil are they doing here, and why don't they make contact with us and get it over with?

Well, of course, there are a few who say they *have* made contact with *them*, but no one wants to believe the stories. Most contact stories show the aliens displaying about the same kind of interest in us as we do in, say, Trobriand Islanders, or Coming of Age in Samoa. Discount the contact reports, and it's still not a bad hypothesis. I know of little to contradict the view that UFO's are mostly filled with graduate students doing a doctoral dissertation on pre-spaceflight cultures.

That explains why they don't make unambiguous contacts. If they did, we wouldn't be a pre-spaceflight culture any more; at the least we'd be going balls out to develop space flight. It also explains the sightings: they don't want to be seen, but once in a while they get careless, as students do.

For that matter, I can envision an extra-terrestrial persuading hiser (no sexists here; they're hermaphroditic) sponsor to let them make "a non-contaminating close encounter. After all, Honored Academician, the observers won't be *believed*! And you've said my thesis isn't original enough, but all the routine observation stuff has been mined out..."

Or what the hell, sometimes they get drunk and put on a show for the primitives, gambling that they won't be reported to their academic superiors. Pasadena knows all about *that* phenomenon: in the areas around Cal Tech the residents shudder as certain times of the year approach.

But blast it all, Pournelle, surely you don't *believe* that?

No. But I wouldn't die of astonishment if it turned out to be true, either. I don't "believe in" flying saucers in the sense that I spend much of my time acting as if Earth were being observed by interstellar graduate students in sophontology; I'm merely inclined to think it's possible, and unable to think of a good alternate theory to account for the UFO observations.

Yes. I know. The extra-terrestrial theory doesn't account for all the observations. On the other hand we don't have any theories that account for all the observations of Martian geography, either, but that doesn't stop me from playing with about half a dozen mutually exclusive hypotheses about Mars.

I do think this. If the "unimaginative" experts of Big Science can take UFO's seriously, while we won't even discuss them, what happens to science fiction writers' reputations as speculators? Not that we need contests on how many impossible things we can believe before breakfast; but can we not at least *speculate* consistent with the observations? Do we, of all people, *dare* ignore UFO's now that study of them is respectable?

* * *

[All the above was written two years before I ever heard of the film Close Encounters; *it has needed no revision and so I offer none. JEP]*

Although Larry Niven and I do a lot of fiction together (MOTE IN GOD'S EYE, INFERNO, LUCIFER'S HAMMER, OATH OF FEALTY, SPIRALS, etc.), we generally don't work together on non-fiction. However, after MOTE IN GOD'S EYE achieved some popularity, Jim Baen, then editor at Galaxy, asked us to do an essay on the "science" in MOTE.

The result, which concludes Part II, is about as far out as you can get.

Building the Mote in God's Eye

by Larry Niven and Jerry Pournelle

Collaborations are unnatural. The writer is a jealous god. He builds his universe without interference. He resents the carping of mentally deficient critics and the editor's capricious demands for revisions. Let two writers try to make one universe, and their defenses get in the way.

But. Our fields of expertise matched each the other's blind spots, unnaturally well. There were books neither of us could write alone. We had to try it.

At first we were too polite, too reluctant to criticize each other's work. That may have saved us from killing each other early on, but it left flaws that had to be torn out of the book later.

We had to build the worlds. From Motie physiognomy we had to build Motie technology and history and life styles. Niven had to be coached in the basic history of Pournelle's thousand-year-old interstellar culture.

It took us three years. At the end we had a novel of

245,000 words . . . which was too long. We cut it to 170,000, to the reader's great benefit. We cut 20,000 words off the beginning, including in one lump our first couple of months of work: a Prologue, a battle between spacegoing warcraft, and a prison camp scene. All of the crucial information had to be embedded in later sections.

We give that Prologue here. When the Moties and the Empire and the star systems and their technologies and philosophies had become one interrelated whole, this is how it looked from New Caledonia system. We called it

MOTELIGHT

Last night at this time he had gone out to look at the stars. Instead a glare of white light like an exploding sun had met him at the door, and when he could see again a flaming mushroom was rising from the cornfields at the edge of the black hemisphere roofing the University. Then had come sound, rumbling, rolling across the fields to shake the house.

Alice had run out in terror, desperate to have her worst fears confirmed, crying, "What are you learning that's worth getting us all killed?"

He'd dismissed her question as typical of an astronomer's wife, but in fact he was learning nothing. The main telescope controls were erratic, and nothing could be done, for the telescope itself was on New Scotland's tiny moon. These nights interplanetary space rippled with the strange lights of war, and the atmosphere glowed with ionization from shock waves, beamed radiation, fusion explosions . . . He had gone back inside without answering.

Now, late in the evening of New Scotland's 27 hour day, Thaddeus Potter, PhD strolled out into the night air.

It was a good night for seeing. Interplanetary war could play hell with the seeing; but tonight the bombardment from New Ireland had ceased. The Imperial Navy had

won a victory.

Potter had paid no attention to the newscasts; still, he appreciated the victory's effects. Perhaps tonight the war wouldn't interfere with his work. He walked thirty paces forward and turned just where the roof of his house wouldn't block the Coal Sack. It was a sight he never tired of.

The Coal Sack was a nebular mass of gas and dust, small as such things go—eight to ten parsecs thick—but dense, and close enough to New Caledonia to block a quarter of the sky. Earth lay somewhere on the other side of it, and so did the Imperial Capital, Sparta, both forever invisible. The Coal Sack hid most of the Empire, but it made a fine velvet backdrop for two close, brilliant stars.

And one of them had changed drastically.

Potter's face changed too. His eyes bugged. His lantern jaw hung loose on its hinges. Stupidly he stared at the sky as if seeing it for the first time.

Then, abruptly, he ran into the house.

Alice came into the bedroom as he was phoning Edwards. "What's happened?" she cried. "Have they pierced the shield?"

"No," Potter snapped over his shoulder. Then, grudgingly, "Something's happened to the Mote."

"Oh for God's sake!" She was genuinely angry, Potter saw. All that fuss about a star, with civilization falling around our ears! But Alice had no love of the stars.

Edwards answered. On the screen he showed naked from the waist up, his long curly hair a tangled bird's nest. "Who the hell—? Thad. I might have known. Thad, do you know what time it is?"

"Yes. Go outside," Potter ordered. "Have a look at the Mote."

"The Mote? The Mote?"

"Yes. It's gone nova!" Potter shouted. Edwards growled, then sudden comprehension struck. He left the screen without hanging up. Potter reached out to dial the *bedroom window transparent. And it was still there.*

Even without the Coal Sack for backdrop Murcheson's Eye would be the brightest object in the sky. At its rising the Coal Sack resembled the silhouette of a hooded man, head and shoulders; and the off-centered red supergiant became a watchful, malevolent eye. The University itself had begun as an observatory founded to study the supergiant.

This eye had a mote: a yellow dwarf companion, smaller and dimmer, and uninteresting. The Universe held plenty of yellow dwarfs.

But tonight the Mote was a brilliant blue-green point. It was almost as bright as Murcheson's Eye itself, and it burned with a purer light. Murcheson's Eye was white with a strong red tinge; but the Mote was blue-green with no compromise, impossibly green.

Edwards came back to the phone. "Thad, that's no nova. It's like nothing ever recorded. Thad, we've got to get to the observatory!"

"I know. I'll meet you there."

"I want to do spectroscopy on it."

"All right."

"God, I hope the seeing holds! Do you think we'll be able to get through today?"

"If you hang up, we'll find out sooner."

"What? Och, aye." Edwards hung up.

* * *

The bombardment started as Potter was boarding his bike. There was a hot streak of light like a very large shooting star; and it didn't burn out, but reached all the way to the horizon. Stratospheric clouds formed and vanished, outlining the shock wave. Light glared on the horizon, then faded gradually.

"Damn," muttered Potter, with feeling. He started the motor. The war was no concern of his, except that he no longer had New Irish students. He even missed some of them. There was one chap from Cohane who . . .

A cluster of stars streaked down in exploding fireworks.

129

Something burned like a new star overhead. The falling stars winked out, but the other light went on and on, changing colors rapidly, even while the shock wave clouds dissipated. Then the night became clear, and Potter saw that it was on the moon.

What could New Ireland be shooting at on New Scotland's moon?

Potter understood then. "You bastards!" he screamed at the sky. "You lousy traitor bastards!"

The single light reddened.

He stormed around the side of Edwards' house shouting, "The traitors bombed the main telescope! Did you see it? All our work—oh."

He had forgotten Edwards' backyard telescope.

It had cost him plenty, and it was very good, although it weighed only four kilograms. It was portable—"Especially," Edwards used to say, "when compared with the main telescope."

He had bought it because the fourth attempt at grinding his own mirror produced another cracked disk and an ultimatum from his now dead wife concerning Number 200 Carbo grains tracked onto her New-Life carpets . . .

Now Edwards moved away from the eyepiece saying, "Nothing much to see there." He was right. There were no features. Potter saw only a uniform aquamarine field.

"But have a look at this," said Edwards. "Move back a bit . . ." He set beneath the eyepiece a large sheet of white paper, then a wedge of clear quartz.

The prism spread a fan-shaped rainbow across the paper. But the rainbow was almost too dim to see, vanishing beside a single line of aquamarine; and that line blazed.

"One line," said Potter. "Monochromatic?"

"I told you yon was no nova."

"Too right it wasn't. But what is it? Laser light? It has to be artificial! Lord, what a technology they've built!"

130

"Och, come now." Edwards interrupted the monologue. "I doubt yon's artificial at all. Too intense." His voice was cheerful. "We're seeing something new. Somehow yon Mote is generating natural coherent light."

"I don't believe it."

Edwards looked annoyed. After all, it was his telescope. "What think you, then? Some booby calling for help? If they were that powerful, they would send a ship. A ship would come thirty-five years sooner!"

"But there's no tramline from the Mote to New Caledonia! Not even theoretically possible. Only link to the Mote has to start inside the Eye. Murcheson looked for it, you know, but he never found it. The Mote's alone out there."

"Och, then how could there be a colony?" Edwards demanded in triumph. "Be reasonable, Thad! We hae a new natural phenomenon, something new in stellar process."

"But if someone is calling—"

"Let's hope not. We could no help them. We couldn't reach them, even if we knew the links! There's no starship in the New Cal system, and there's no likely to be until the war's over." Edwards looked up at the sky. The moon was a small, irregular half-disk; and a circular crater still burned red in the dark half.

A brilliant violet streak flamed high overhead. The violet light grew more intense and flared white, then vanished. A warship had died out there.

"Ah, well," Edwards said. His voice softened. "If someone's calling he picked a hell of a time for it. But at least we can search for modulations. If the beam is no modulated, you'll admit there's nobody there, will you not?"

"Of course," said Potter.

* * *

In 2862 there were no starships behind the Coal Sack. On the other side, around Crucis and the Capital, a tiny fleet still rode the force paths between stars to the worlds

Sparta controlled. There were fewer loyal ships and worlds each year.

The summer of 2862 was lean for New Scotland. Day after day a few men crept outside the black dome that defended the city; but they always returned at night. Few saw the rising of the Coal Sack.

It climbed weirdly, its resemblance to a shrouded human silhouette marred by the festive two-colored eye. The Mote burned as brightly as Murcheson's Eye now. But who would listen to Potter and Edwards and their crazy tales about the Mote? The night sky was a battlefield, dangerous to look upon.

The war was not really fought for the Empire now. In the New Caledonia system the war continued because it would not end. Loyalist and Rebel were meaningless terms; but it hardly mattered while bombs and wrecked ships fell from the skies.

Henry Morrissey was still head of the University Astronomy Department. He tried to talk Potter and Edwards into returning to the protection of the Langston Field. His only success was that Potter sent his wife and two sons back with Morrissey. Edwards had no living dependents, and both refused to budge.

Morrissey was willing to believe that something had happened to the Mote, but not that it was visible to the naked eye. Potter was known for his monomaniacal enthusiasms.

The Department could supply them with equipment. It was makeshift, but it should have done the job. There was laser light coming from the Mote. It came with terrific force, and must have required terrific power, and enormous sophistication to build that power. No one would build such a thing except to send a message.

And there was no message. The beam was not modulated. It did not change color, or blink off and on, or change in intensity. It was a steady, beautifully pure, terribly intense beam of coherent light.

Potter watched to see if it might change silhouette, staring for hours into the telescope. Edwards was no help at all. He alternated between polite gloating at having proved his point, and impolite words muttered as he tried to investigate the new "stellar process" with inadequate equipment. The only thing they agreed on was the need to publish their observations, and the impossibility of doing so.

One night a missile exploded against the edge of the black dome. The Langston Field protecting University City could only absorb so much energy before radiating inward, vaporizing the town, and it took time to dissipate the hellish fury poured into it. Frantic engineers worked to radiate away the shield energy before the generators melted to slag.

They succeeded, but there was a burn-through: a generator left yellow-hot and runny. A relay snapped open, and New Caledonia stood undefended against a hostile sky. Before the Navy could restore the Field a million people had watched the rising of the Coal Sack.

"I came to apologize," Morrissey told Potter the next morning. "Something damned strange has happened to the Mote. What have you got?"

He listened to Potter and Edwards, and he stopped their fight. Now that they had an audience they almost came to blows. Morrissey promised them more equipment and retreated under the restored shield. He had been an astronomer in his time. Somehow he got them what they needed.

Weeks became months. The war continued, wearing New Scotland down, exhausting her resources. Potter and Edwards worked on, learning nothing, fighting with each other and screaming curses at the New Irish traitors.

They might as well have stayed under the shield. The Mote produced coherent light of amazing purity. Four months after it began the light jumped in intensity and stayed that way. Five months later it jumped again.

It jumped once more, four months later, but Potter and

133

Edwards didn't see it. That was the night a ship from New Ireland fell from the sky, its shield blazing violet with friction. It was low when the shield overloaded and collapsed, releasing stored energy in one ferocious blast.

Gammas and photons washed across the plains beyond the city, and Potter and Edwards were carried into the University hospital by worried students. Potter died three days later. Edwards walked for the rest of his life with a backpack attached to his shoulders: a portable life support system.

*　*　*

It was 2870 on every world where clocks still ran when the miracle came to New Scotland.

An interstellar trading ship, long converted for war and recently damaged, fell into the system with her Langston Field intact and her hold filled with torpedoes. She was killed in the final battle, but the insurrection on New Ireland died as well. Now all the New Caledonia system was loyal to the Empire; and the Empire no longer existed.

The University came out from under the shield. Some had forgotten that the Mote had once been a small yellow-white point. Most didn't care. There was a world to be tamed, and that world had been bare rock terraformed in the first place. The fragile imported biosphere was nearly destroyed, and it took all their ingenuity and work to keep New Scotland inhabitable.

They succeeded because they had to. There were no ships to take survivors anywhere else. The Yards had been destroyed in the war, and there would be no more interstellar craft. They were alone behind the Coal Sack.

The Mote continued to grow brighter as the years passed. Soon it was more brilliant than the Eye; but there were no astronomers on New Scotland to care. In 2891 the Coal Sack was a black silhouette of a hooded man. It had one terribly bright blue-green eye, with a red fleck in it.

*　*　*

One night at the rising of the Coal Sack, a farmer named

134

Howard Grote Littlemead was struck with inspiration. It came to him that the Coal Sack was God, and that he ought to tell someone.

Tradition had it that the Face of God could be seen from New Caledonia; and Littlemead had a powerful voice. Despite the opposition of the Imperial Orthodox Church, despite the protests of the Viceroy and the scorn of the University staff, the Church of Him spread until it was a power on New Scotland.

It was never large, but its members were fanatics; and they had the miracle of the Mote, which no scientist could explain. By 2895 the Church of Him was a power among New Scot farmers, but not in the cities. Still, half the population worked in the fields. The converter kitchens had all broken down.

By 2900 New Scotland had two working interplanetary spacecraft, one of which could not land. Its Langston Field had died. The term was appropriate. When a piece of Empire technology stopped working, it was dead. It could not be repaired. New Scotland was becoming primitive.

For forty years the Mote had grown. Children refused to believe that it had once been called the Mote. Adults knew it was true, but couldn't remember why. They called the twin stars Murcheson's Eye, and believed that the red supergiant had no special name.

The records might have showed differently, but the University records were suspect. The Library had been scrambled by electromagnetic pulses during the years of siege. It had large areas of amnesia.

In 2902 the Mote went out.

Its green light dimmed to nothing over a period of several hours; but that happened on the other side of the world. When the Coal Sack rose above University City that night, it rose as a blinded man.

All but a few remnants of the Church of Him died that year. With the aid of a handful of sleeping pills Howard Grote Littlemead hastened to meet his God . . . possibly to demand an explanation.

Astronomy also died. There were few enough astronomers and fewer tools; and when nobody could explain the vanishing of the Mote . . . and when telescopes turned on the Mote's remnant showed only a yellow dwarf star, with nothing remarkable about it at all . . .

People stopped considering the stars. They had a world to save.

The Mote was a G-2 yellow dwarf, thirty-five light years distant: a white point at the edge of Murcheson's Eye. So it was for more than a century, while the Second Empire rose from Sparta and came again to New Caledonia.

Then astronomers read old and incomplete records, and resumed their study of the red supergiant known as Murcheson's Eye; but they hardly noticed the Mote.

And the Mote did nothing unusual for one hundred and fifteen years.

* * *

Thirty-five light years away, the aliens of Mote Prime had launched a light-sail spacecraft, using batteries of laser cannon powerful enough to outshine a neighboring red supergiant.

As for why they did it that way, and why it looked like that, and what the bejeesus is going on . . . explanations follow.

* * *

Most hard science fiction writers follow standard rules for building worlds. We have formulae and tables for getting the orbits right, selecting suns of proper brightness, determining temperatures and climates, building a plausible ecology. Building worlds requires imagination, but a lot of the work is mechanical. Once the mechanical work is done the world may suggest a story, or it may even design its own inhabitants. Larry Niven's "known space" stories include worlds which have strongly affected their colonists. Or the exceptions to the rules may form stories. Why does Mote Prime, a nominally Earthlike world, remind so many people of the planet Mars? What strangeness in its evolution made the atmosphere so helium-rich? This goes beyond mechanics.

In THE MOTE IN GOD'S EYE (Simon and Schuster, 1974) we built not only worlds, but cultures.

From the start MOTE was to be a novel of first contact. After our initial story conference we had larger ambitions: MOTE would be, if we could write it, the *epitome* of first contact novels. We intended to explore every important problem arising from first contact with aliens—and to look at those problems from both human and alien viewpoints.

That meant creating cultures in far more detail than is needed for most novels. It's easy, when a novel is heavy with detail, for the details to get out of hand, creating glaring inconsistencies. (If civilization uses hydrogen fusion power at such a rate that world sea level has dropped by two feet, you will not have people sleeping in abandoned movie houses.) To avoid such inconsistencies we worked a great deal harder developing the basic technologies of both the Motie (alien) and the human civilizations.

In fact, when we finished the book we had nearly as much unpublished material as ended up in the book. There are many pages of data on Motie biology and evolutionary history; details on Empire science and technology; descriptions of space battles, how worlds are terraformed, how light-sails are constructed; and although these background details affected the novel and dictated what we would actually write, most of them never appear in the book.

We made several boundary decisions. One was to employ the Second Empire period of Pournelle's future history. That Empire existed as a series of sketches with a loose outline of its history, most of it previously published. MOTE had to be consistent with the published material.

Another parameter was the physical description of the

aliens. Incredibly, that's all we began with: a detailed description of what became the prototype Motie, the Engineer: an attempt to build a nonsymmetrical alien, left over from a Niven story that never quite jelled. The history, biology, evolution, sociology, and culture of the Moties were extrapolated from that being's shape during endless coffee-and-brandy sessions.

That was our second forced choice. The Moties lived within the heart of the Empire, but had never been discovered. A simple explanation might have been to make the aliens a young civilization just discovering space travel, but that assumption contradicted Motie history as extrapolated from their appearance. We found another explanation in the nature of the Alderson Drive.

EMPIRE TECHNOLOGY

The most important technological features of the Empire were previously published in other stories: the Alderson Drive and Langston Field.

Both were invented to Jerry Pournelle's specifications by Dan Alderson, a resident genius at Cal Tech's Jet Propulsion Laboratories. It had always been obvious that the Drive and Field would affect the cultures that used them, but until we got to work on MOTE it wasn't obvious just how profound the effects would be.

The Alderson Drive

Every SF writer eventually must face the problem of interstellar transportation. There are a number of approaches. One is to deny faster-than-light travel. This in practice forbids organized interstellar civilizations.

A second approach is to ignore General and Special Relativity. Readers usually won't accept this. It's a cop-out, and except in the kind of story that's more allegory than science fiction, it's not appropriate.

Another method is to retreat into doubletalk about hyperspace. Doubletalk drives are common enough. The problem is that when everything is permitted, nothing is forbidden. Good stories are made when there are difficulties to overcome, and if there are no limits to "hyperspace travel" there are no real limits to what the heroes and villains can do. In a single work the "difficulties" can be planned as the story goes along, and the drive then redesigned in rewrite; but we couldn't do that here.

Our method was to work out the Drive in detail and live with the resulting limitations. As it happens, the limits on the Drive influenced the final outcome of the story; but they were not invented for that purpose.

The Alderson Drive is consistent with everything presently known about physics. It merely assumes that additional discoveries will be made in about thirty years, at Cal Tech (as a tip o' the hat to Dan Alderson). The key event is the detection of a "fifth force."

There are four known forces in modern physics: two sub-nuclear forces responsible respectively for alpha and beta decay; electromagnetism, which includes light; and gravity. The Alderson force, then, is the fifth, and it is generated by thermonuclear reactions.

The force has little effect in our universe; in fact, it is barely detectable. Simultaneously with the discovery of the fifth force, however, we postulate the discovery of a second universe in point-to-point congruence with our own. The "continuum universe" differs from the one we're used to in that there are no known quantum effects there.

Within that universe particles may travel as fast as they can be accelerated; and the fifth force exists to accelerate them.

There's a lot more, including a page or so of differential equations, but that's the general idea.

You can get from one universe to another. For every construct in our universe there can be created a "correspondence particle" in the continuum universe. In order for your construct to go into and emerge from the continuum universe without change you must have some complex machinery to hold everything together and prevent your ship—and crew—from being disorganized into elementary particles.

Correspondence particles can be boosted to speeds faster than light: in fact, to speeds nearly infinite as we measure them. Of course they cannot emerge into our universe at such speeds: they have to lose their energy to emerge at all. More on that in a moment.

There are severe conditions to entering and leaving the continuum universe. To emerge from the continuum universe you must exit with precisely the same potential energy (measured in terms of the fifth force, not gravity) as you entered. You must also have zero kinetic energy relative to a complex set of coordinates that we won't discuss here.

The fifth force is created by thermonuclear reactions: generally, that is, in stars. You may travel by using it, but only along precisely defined lines of equipotential flux: tramways or tramlines.

Imagine the universe as a thin rubber sheet, very flat. Now drop heavy rocks of different weights onto it. The rocks will distort the sheet, making little cone-shaped (more or less) dimples. Now put two rocks reasonably close together: the dimples will intersect in a valley. The intersection will have a "pass," a region higher than the low points where the rocks (stars) lie, but lower than the general level of the rubber sheet.

The route from one star to another through that "pass" is the tramline. Possible tramlines lie between each two stars, but they don't always exist, because when you add third and fourth stars to the system they may interfere, so there is no unique gradient line. If this seems confusing, don't spend a lot of time worrying about it; we'll get to the effects of all this in a moment.

You may also imagine stars to be like hills; move another star close and the hills will intersect. Again, from summit to summit there will be one and only one line that preserves the maximum potential energy for that level. Release a marble on one hill and it will roll down, across the saddle, and up the side of the other. That too is a tramline effect. It's generally easier to think of the system as valleys rather than hills, because to travel from star to star you have to get over that "hump" between the two. The fifth force provides the energy for that.

You enter from the quantum universe. When you travel in the continuum universe you continually lose kinetic energy; it "leaks." This can be detected in our universe as photons. The effect can be important during a space battle. We cut such a space battle from MOTE, but it still exists, and we may yet publish it as a novella.

To get from the quantum to the continuum universe you must supply power, and this is available only in quantum terms. When you do this you turn yourself into a correspondence particle; go across the tramline; and come out at the point on the other side where your potential energy is equal to what you entered with, plus zero kinetic energy (in terms of the fifth force and complex reference axes).

For those bored by the last few paragraphs, take heart: we'll leave the technical details and get on with what it all means.

Travel by Alderson Drive consists of getting to the proper Alderson Point and turning on the Drive. Energy is used. You vanish, to reappear in an immeasurably short time at the Alderson Point in another star system some several light years away. If you haven't done everything right, or aren't at the Alderson Point, you turn on your drive and a lot of energy vanishes. You don't move. (In fact you do move, but you instantaneously reappear in the spot where you started.)

That's all there is to the Drive, but it dictates the structure of an interstellar civilization.

To begin with, the Drive works only from point to point across interstellar distances. Once in a star system you must rely on reaction drives to get around. There's no magic way from, say, Saturn to Earth: you've got to slog across.

Thus space battles are possible, and you can't escape battle by vanishing into hyperspace, as you could in future history series such as Beam Piper's and Gordon Dickson's. To reach a given planet you must travel across its stellar system, and you must enter that system at one of the Alderson Points. There won't be more than five or six possible points of entry, and there may only be one.

Star systems and planets can be thought of as continents and islands, then, and Alderson Points as narrow sea gates such as Suez, Gibraltar, Panama, Malay Straits, etc. To carry the analogy further, there's telegraph but no radio: the fastest message between star systems is one carried by a ship, but within star systems messages go much faster than the ships . . .

Hmm. This sounds a bit like the early days of steam. NOT sail; the ships require fuel and sophisticated repair facilities. They won't pull into some deserted star system and rebuild themselves unless they've carried the spare parts along. However, if you think of naval actions in the periods between the Crimean War and World War One, you'll have a fair picture of conditions as implied by the Alderson Drive.

The Drive's limits mean that uninteresting stellar systems won't be explored. There are too many of them. They may be used as crossing-points if the stars are conveniently placed, but stars not along a travel route may never be visited.

Reaching the Mote, or leaving it, would be damned inconvenient. Its only tramline reaches to a star only a third of a light year away—Murcheson's Eye, the red supergiant —and ends deep inside the red-hot outer envelope. The aliens' only access to the Empire is across thirty-five light years of interstellar space—which no Empire ship would ever see. The gaps between the stars are as mysterious to the Empire as they are to you.

* * *

Langston Field

Our second key technological building block was the Langston Field, which absorbs and stores energy in proportion to the fourth power of incoming particle energy: that is, a slow-moving object can penetrate it, but the faster it's moving (or hotter it is) the more readily it is absorbed.

(In fact it's not a simple fourth-power equation; but perhaps you don't need third-order differential equations for amusement.)

The Field can be used for protection against lasers, thermonuclear weapons, and nearly anything else. It isn't a perfect defense, however. The natural shape of the Field is a solid. Thus it wants to collapse and vaporize everything inside it. It takes energy to maintain a hole inside the Field, and more energy to open a control in it so that you can cause it selectively to radiate away stored energy. You don't get something for nothing.

This means that if a Field is overloaded, the ship inside vanishes into vapor. In addition, *parts* of the Field can be momentarily overloaded: a sufficiently high energy impacting a small enough area will cause a temporary Field collapse, and a burst of energy penetrates to the inside. This can damage a ship without destroying it.

Cosmography

We've got to invent a term. What is a good word to mean the equivalent of "geography" as projected into interstellar space? True, planetologists have now adopted "geology" to mean geophysical sciences applied to any planet, not merely Earth; and one might reasonably expect "geography" to be applied to the study of physical features of other planets—but we're concerned here with the relationship of star systems to each other.

We suggest cosmography, but perhaps that's too broad? Should that term be used for relationships of *galaxies*, and mere star system patterns be studied as "astrography"? After all, "astrogator" is a widely used term meaning "navigator" for interstellar flight.

Some of the astrography of MOTE was given because it had been previously published. In particular, the New Caledonia system, and the red supergiant known as Murcheson's Eye, had already been worked out. There were also published references to the history of New Caledonia.

We needed a red supergiant in the Empire. There's only one logical place for that, and previously published stories had placed one there: Murcheson's Eye, behind the Coal Sack. It *has* to be behind the Coal Sack: if there were a supergiant that close anywhere else, we'd see it now.

Since we had to use Murcheson's Eye, we had to use New Caledonia. Not that this was any great imposition: New Scotland and New Ireland are interesting places, terraformed planets, with interesting features and interesting cultures.

There was one problem, though: New Scotland is inhabited by New Scots, a people who have preserved their sub-culture for a long time and defend it proudly. Thus, since much of the action takes place on New Scotland, some of the characters, including at least one major character, *had* to be New Scot. For structural reasons we had only two choices: the First Officer or the Chief Engineer.

We chose the Chief Engineer, largely because in the contemporary world it is a fact that a vastly disproportionate number of ship's engineers are Scots, and that seemed a reasonable thing to project into the future.

Alas, some critics have resented that, and a few have accused us of stealing Mr. Sinclair from *Star Trek*. We didn't. Mr. Sinclair is what he is for perfectly sound astrographical reasons.

The astrography eventually dictated the title of the book. Since most of the action takes place very near the Coal Sack we needed to know how the Coal Sack would look close up from the back side. Eventually we put swirls of interplanetary dust in it, and evolving proto-stars, and all manner of marvels; but those came after we got *very* close. The first problem was the Coal Sack seen from ten parsecs.

Larry Niven hit on the happy image of a hooded man, with the super-giant where one eye might be. The super-giant has a small companion, a yellow dwarf not very different from our Sun. If the supergiant is an eye—Murcheson's Eye—then the dwarf is, of course, a mote in that eye.

But if the Hooded Man is seen by backward and superstitious peoples as the Face of God ... then the name for the Mote becomes inevitable ... and once suggested, The Mote In God's Eye is a near irresistable title. (Although in fact Larry Niven did resist it, and wanted "The Mote In Murcheson's Eye" up to the moment when the publisher argued strongly for the present title ...)

The Ships

Long ago we acquired a commercial model called "The Explorer Ship Leif Ericson," a plastic spaceship of intriguing design. It is shaped something like a flattened pint whiskey bottle with a long neck. The "Leif Ericson," alas, was killed by general lack of interest in spacecraft by model buyers; a ghost of it is still marketed in hideous glow-in-the-dark color as some kind of flying saucer.

It's often easier to take a detailed construct and work within its limits than it is to have too much flexibility. For fun we tried to make the Leif Ericson work as a model for an Empire naval vessel. The exercise proved instructive.

First, the model is of a *big* ship, too poorly designed in shape ever to be carried aboard another vessel. Second, it had fins. Fins are only useful for atmosphere flight: what purpose would be served in having atmosphere capabilities on a large ship?

This dictated the class of ship: it must be a cruiser or battlecruiser. Battleships and dreadnaughts wouldn't ever land, and would be cylindrical or spherical to reduce surface area. Our ship was too large to be a destroyer (an expendable ship almost never employed on missions except as part of a flotilla). Cruisers and battlecruisers can be sent on independent missions.

MacArthur, a General Class Battlecruiser, began to emerge. She can enter atmosphere, but rarely does so, except when long independent assignments force her to seek fuel on her own. She can do this in either of two ways: go to a supply source, or fly into the hydrogen-rich atmosphere of a gas giant and scoop. There were scoops on the model, as it happens.

She has a large pair of doors in her hull, and a spacious compartment inside: obviously a hangar deck for carrying auxiliary craft. Hangar deck is also the only large compartment in her, and therefore would be the normal place of assembly for the crew when she isn't under battle conditions.

The tower on the model looked useless, and was almost ignored, until it occurred to us that on long missions not under acceleration it would be useful to have a high-gravity area. The ship is a bit thin to have much gravity in the "neck" without spinning her far more rapidly than you'd like; but with the tower, the forward area gets normal gravity without excessive spin rates.

And on, and so forth. In the novel, *Lenin* was designed from scratch; and of course we did have to make some modifications in Leif Ericson before she could become INSS *MacArthur*; but it's surprising just how much detail you can work up through having to live with the limits of a model . . .

SOCIOLOGY
The Alderson Drive and the Langston Field determine what kinds of interstellar organizations will be possible. There will be alternatives, but they have to fit into the limits these technologies impose.

In THE MOTE IN GOD'S EYE we chose Imperial Aristocracy as the main form of human government. We've been praised for this: Dick Brass in a *New York Post* review concludes that we couldn't have chosen anything else, and other critics have applauded us for showing what such a society might be like.

Fortunately there are no Sacred Cows in science fiction. Maybe we should have stuck to incest? Because other critics have been horrified! Do we, they ask, really *believe* in imperial government? and *monarchy*?

That depends on what they mean by "believe in." Do we think it's desirable? We don't have to say. Inevitable? Of course not. Do we think it's *possible*? Damn straight.

The political science in MOTE is taken from C. Northcote Parkinson's EVOLUTION OF POLITICAL THOUGHT, Parkinson himself echoes Aristotle.

It is fashionable to view history as a linear progression: things get better, never worse, and of course we'll never go back to the bad old days of (for instance) personal government. Oddly enough, even critics who have complained about the aristocratic pyramid in MOTE—and thus rejected our Empire as absurd—have been heard to complain about "Imperial Presidency" in the USA. How many readers would bet long odds against John-John Kennedy becoming President within our lifetimes?

Any pretended "science" of history is the bunk. That's the problem with Marxism. Yet Marx wrote a reasonable economic view of history up to his time, and some of his principles may be valid.

Military history is another valid way to view the last several thousand years—but no one in his right mind would pretend that a history of battles and strategies is the whole of the human story. You may write history in terms of medical science, in terms of rats, lice, and plagues, in terms of agricultural development, in terms of strong leadership personalities, and each view will hold some truth.

There are many ways to view history, and Aristotle's cycles as brought up to date by Parkinson make one of the better ones. For those who don't accept that proposition, we urge you at least to read Parkinson before making up your minds and closing the door.

The human society in MOTE is colored by technology and historical evolution. In MOTE's future history the United States and the Soviet Union form an alliance and together dominate the world during the last decades of the 20th Century. The alliance doesn't end their rivalry, and doesn't make the rulers or people of either nation love their partners.

The CoDominium Alliance needs a military force. Military people need something or someone they can give loyalty; few men ever risked their lives for a "standard of living" and there's nothing stupider than dying for one's standard of living—unless it's dying for someone else's.

Do the attitudes of contemporary police and soldiers lead us to suppose that "democracy" or "the people" inspire loyalty? The proposition is at least open to question. In the future that leads to MOTE, a Russian Admiral named Lermontov becomes leader of CoDominium forces, and although he is not himself interested in founding a dynasty, he transfers the loyalty of the Fleet to leaders who are.

He brings with him the military people at a time of great crisis. Crises have often produced strong loyalties to single leaders: Churchill, Roosevelt, George Washington, John F. Kennedy during the Cuban Crisis, etc. (A year after Kennedy's death Senator Pastore could address a national convention and get standing ovations with the words "There stood John Kennedy, TEN FEET TALL!!!")

Thus develops the Empire.

Look at another trend: personal dictatorship. There are as many people ruled by tyrants as by "democracy" in 1979, and even in the democracies charges of tyranny are not lacking. Dictatorships may not be the wave of the future—but is it unreasonable to suppose they might be?

Dictatorship is often tried in times of severe crisis: energy crisis, population crisis, pollution crisis, agricultural crisis

—surely we do not lack for crises? The trouble with dictatorship is that it itself generates a succession crisis when the old man bows out. Portugal seems to be going through such at this moment. Chile, Uganda, Brazil, name your own examples: anyone want to bet that some of these won't turn to a new Caudillo with relief?

How to avoid succession crisis? One traditional method is to turn Bonapartist: give the job to a relative or descendant of the dictator. He may not do the job very well, but after enough crises people are often uninterested in whether the land is governed well. They just want things *settled* so they can get on with everyday life.

Suppose the dictator's son does govern well? A new dynasty is founded, and the trappings of legitimacy are thrust onto the new royal family. To be sure, the title of "King" may be abandoned. Napoleon chose to be "Emperor of the French," Cromwell chose "Lord Protector," and we suppose the US will be ruled by Presidents for a long time—but the nature of the Presidency, and the way one gets the office, may change.

See, for example, Niven's use of "Secretary-General" in the tales of Svetz the time-traveller.

We had a choice in MOTE: to keep the titles as well as the structure of aristocratic empire, or abandon the titles and retain the structure only. We could have abolished "Emperor" in favor of "President," or "Chairperson," or "Leader," or "Admiral," or "Posnitch." The latter, by the way, is the name of a particularly important President honored for all time by having his name adopted as the title for Leader...

We might have employed titles other than Duke (originally meant "leader" anyway) and Count (Companion to the king) and Marquis (Count of the frontier marches). Perhaps we should have. But any titles used would have been *translations* of whatever was current in the time of the novel, and the traditional titles had the effect of letting the reader know quickly the approximate status and some of the duties of the characters.

There are hints all through MOTE that the structure of government is not a mere carbon copy of the British Empire or Rome or England in the time of William III. On the other hand there are similarities, which are forced onto the Empire by the technology we assumed.

Imperial government is not inevitable. It is possible.

The alternate proposition is that we of 1979 are so advanced that we will never go back to the bad old days. Yet we can show you essays "proving" exactly that proposition —and written thousands of years ago. There's a flurry of them every few centuries.

We aren't the first people to think we've "gone beyond" personal government, personal loyalties, and a state religion. Maybe we won't be the last.

Anyway, MOTE is supposed to be entertainment, not an essay on the influence of science on social organization. (You're getting *that* here.)

The Empire is what it is largely because of the Alderson Drive and Langston Field. Without the Drive an Empire could not form. Certainly an interstellar Empire would look very different if it had to depend on lightspeed messages to send directives and receive reports. Punitive expeditions would be nearly impossible, hideously expensive, and probably futile: you'd be punishing the grandchildren of a generation that seceded from the Empire, or even a planet that put down the traitors after the message went out.

Even a rescue expedition might never reach a colony in trouble. A coalition of bureaucrats could always collect the funds for such an expedition, sign papers certifying that the ships are on the way, and pocket the money . . . in sixty years someone might realize what had happened, or not.

The Langston Field is crucial to the Empire, too. Naval vessels can survive partial destruction and keep fighting. Ships carry black boxes—plug-in sets of spare parts—and large crews who have little to do unless half of them get killed. That's much like the navies of fifty years ago.

A merchant ship might have a crew of forty. A warship of similar size carries a crew ten times as large. Most have little to do for most of the life of the ship. It's only in battles that the large number of self-programming computers become important. *Then* the outcome of the battle may depend on having the largest and best-trained crew—and there aren't many prizes for second place in battle.

Big crews with little to do demand an organization geared to that kind of activity. Navies have been doing that for a long time, and have evolved a structure that they tenaciously hold on to.

Without the Field as defense against lasers and nuclear weapons, battles would become no more than offensive contests. They'd last microseconds, not hours. Ships would be destroyed or not, but hardly ever wounded. Crews would tend to be small, ships would be different, including something like the present-day aircraft carriers. Thus technology dictates Naval organization.

It dictates politics, too. If you can't get the populace, or a large part of it, under a city-sized Field, then any given planet lies naked to space.

If the Drive allowed ships to sneak up on planets, materializing without warning out of hyperspace, there could be no Empire even with the Field. There'd be no Empire because belonging to an Empire wouldn't protect you. Instead there might be populations of planet-bound serfs ruled at random by successive hordes of space pirates. Upward mobility in society would consist of getting your own ship and turning pirate.

Given Drive and Field, though, Empires are possible. What's more likely? A representative confederacy? It would hardly inspire the loyalty of the military forces, whatever else it might do. (In the War Between the States, the Confederacy's main problem was that the troops were loyal to their own State, not the central government.)

151

Each stellar system independent? That's reasonable, but is it stable? Surely there might be pressures toward unification of at least parts of interstellar space.

How has unification been achieved in the past? Nearly always by conquest or colonization or both. How have they been held together? Nearly always by loyalty to a leader, an Emperor, or a dynasty, generally buttressed by the trappings of religion and piety. Even Freethinkers of the last century weren't ashamed to profess loyalty to the Widow of Windsor ...

Government over large areas needs emotional ties. It also needs *stability*. Government by 50%-plus-one hasn't enjoyed particularly stable politics—and it lasts only so long as the 50%-minus-one minority is willing to submit. Is heredity a rational way to choose leaders? It has this in its favor: the leader is known from an early age to be destined to rule, and can be educated to the job. Is that preferable to education based on how to *get* the job? Are elected officials better at governing, or at winning elections?

Well, at least the counter-case can be made. That's all we intended to do. We chose a stage of Empire in which the aristocracy was young and growing and dynamic, rather than static and decadent; when the aristocrats are more concerned with duty than with privilege; and we made no hint that we thought that stage would last forever.

RANDOM DETAILS

Robert Heinlein once wrote that the best way to give the flavor of the future is to drop in, without warning, some strange detail. He gives as an example, "The door dilated."

We have a number of such details in MOTE. We won't spoil the book by dragging them all out in a row. One of the most obvious we use is the personal computer, which not only does computations, but also puts the owner in contact with any near-by data bank; in effect it will give the answer to any question whose answer is known and that you think to ask.

Thus no idiot block gimmicks in MOTE. Our characters may fail to guess something, or not put information together in the right way, but they won't *forget* anything important. The closest that comes to happening is when Sally Fowler can't quite remember where she filed the tape of a conversation, and she doesn't take long to find it then.

On the other hand, people can be swamped with too much information, and that does happen.

There were many other details, all needed to keep the story moving. A rational kind of space suit, certainly different from the clumsy things used now. Personal weapons. The crystal used in a banquet aboard *MacArthur*: crystal strong as steel, cut from the windshield of a wrecked First Empire reentry vehicle, indicating the higher technology lost in that particular war. Clothing and fashion; the status of women; myriads of details of everyday life.

Not that *all* of these differ from the present. Some of the things we kept the same probably will change in a thousand years. Others ... well, the customs associated with wines and hard liquors are old and stable. If we'd changed everything, and made an attempt to portray every detail of our thousand-year-advanced future, the story would have gotten bogged down in details.

MOTE is probably the only novel ever to have a planet's orbit changed to save a line.

New Chicago, as it appeared in the opening scenes of the first draft of MOTE, was a cold place, orbiting far from its star. It was never a very important point, and Larry Niven didn't even notice it.

Thus when he introduced Lady Sandra Liddell Leonovna Bright Fowler, he used as viewpoint character a Marine guard sweating in hot sunlight. The Marine thinks, "She doesn't sweat. She was carved from ice by the finest sculptor that ever lived."

Now that's a good line. Unfortunately it implies a hot

planet. If the line must be kept, the planet must be moved.

So Jerry Pournelle moved it. New Chicago became a world much closer to a somewhat cooler sun. Its year changed, its climate changed, its whole history had to be changed . . .

Worth it, though. Sometimes it's easier to build new worlds than think up good lines . . .

PART THREE:

A STEP FARTHER IN: BLACK HOLES

COMMENTARY

I was one of the first science fiction writers to use black holes in a story. Not *the* first, I hasten to add; but I did make the first use of gravitation waves, and therein lies a tale.

I had just published my story "He Fell Into a Dark Hole" and was in fact reading it—writers generally do read their own work the first time they see it in print—when the telephone rang. The caller was Dr. Robert Forward of the Hughes Research Laboratories; Forward, I later learned, is one of the foremost authorities on gravitation, and holds patents on gadgetry such as a "mass detector."

He had been preparing a paper on gravitation waves and how they might prove to be dangerous; that was the theme of my story; and I had beaten him into print. So not only did I make first use of them in science fiction, but my story may have the first publication anywhere to draw attention to certain aspects of gravity waves.

The result is that I am, to many fans, a sort of "proprietor" of black holes; a role I'm willing to fill for a while. Which gives me the right to tell you about them, as this section does.

Gravity Waves, Black Holes, and Cosmic Censors

I suppose most readers are at least partly aware of the ongoing research on detection of gravitational waves, but it does no harm to summarize a bit. In the Newtonian universe, gravity is a "force" that acts through a field; that is, although it is 10 times weaker than electromagnetism, it's not fundamentally different.

This holds true in the realm of special relativity also: special relativity is the theory that asserts that no material object, and no signal, can travel faster than light. There's a lot of evidence for special relativity, and no really good counter theory lurks in the wings to take its place.

The general theory of relativity is another breed of cat entirely. There are several contenders in that realm, and experimental evidence offers no clearcut way to choose one or another. General relativity does away with gravity fields altogether: in that theory, gravity results from the geometry of space, and is not a "force" at all.

(That is: mass—or energy for that matter—distorts space, bending it; and it is this curvature of the fabric of space itself that causes the effects we call "gravity.")

Whether gravity fields "exist" or merely result from geometry, nearly all theorists believe gravitational attraction propagates with the speed of light. If matter is created —or destroyed—the rest of the universe won't be instantly affected, but must wait until the gravitational effect, travelling at light speed, reaches it.

Thus "gravitation waves," which will have a frequency and an amplitude much like light, but which may also have some rather strange properties as well.

In theory, if we could detect and examine gravitational waves, we might be able to tell whether they result from a field and are thus similar to magnetism, or if they are merely a property of space and its geometry. Unfortunately, gravity is an incredibly weak force. It requires the mass of the whole earth merely to pull things with a puny 980 cm/sec^2 acceleration—and we can overcome that with rather small magnets, or chemical rockets, or even our own muscles when we jump.

Because gravity is so weak, it's hard to play with. You can't turn on a "gravity wave generator" and fiddle with the resulting forces to see if they refract, or can be tuned, or whatever. You can't wiggle a mass to generate gravity waves, because you can't get a large enough mass held into place to be wiggled. It's not even possible to blow off an atomic weapon, turning some matter into energy, and measure the effect of the matter vanishing; the effect is just too small to be noticed, and it's hidden among the rather drastic side effects.

However, there are a number of theoretical ways that gravity waves might be generated by the universe: stars collapsing into black holes or neutronium would do it, for example. The universe might be riddled with gravitational waves, but they'd be terribly weak, and require delicate and sophisticated apparatus to detect them.

Some years ago, Dr. Joseph Weber of the University of Maryland decided to build a gravity wave antenna. He took a large aluminum cylinder and covered it with strain gauges. The idea was that so long as the cylinder was acted on only by the steady gravity of Earth, it would be in a stable configuration; but if a gravity wave passed through it, the cylinder would be distorted, and the strain gauges would show it.

He had to compensate for temperature, and isolate it from vibration, and worry about a lot of other things, but the technology had been developed: the antenna was built. It was incredibly sensitive, able to detect distortions on the order of an atomic diameter. It was also able to detect student demonstrations outside the library, trucks rumbling along the highway a mile distant, and other unwanted events.

The solution to the latter problem was simple: build another copy of the antenna and place it 1000 kilometers away; now hook the two together, and pay no attention to any event that doesn't affect both. Such "coincidences" should be due to a force affecting both antennae—earthquakes take time to propagate and their effects move much slower than lightspeed—the output should be reliable.

Unfortunately, it isn't as straightforward as that. The instruments must be very sensitive, and thus there's a lot of chatter from them. By the laws of chance, some of this chatter will be simultaneous, or near enough so, and thus you are guaranteed some false positive results. The output of the gravity wave detectors, therefore, needs careful analysis to decide what's real data and what's chance.

Weber immediately got results. He got a lot of results, far too many for chance. Unfortunately, there were far too many for cosmologists to believe. As a result of Weber's early reports, some cosmologists estimated that as much as 98% of the universe must be inside black holes.

The argument went this way: something is producing gravity waves. We can't see enough matter to account for the events, but normal matter falling into a black hole would produce gravity waves. Therefore—

There were other cosmologists who wanted to believe this for different reasons. Readers familiar with black holes must excuse me: it's now necessary to discuss their basics for a moment.

* * *

A black hole is a theoretical construct that can be derived from both general relativity and the older Newtonian universe; in fact, the first speculations about black holes come from Laplace back in 1798. If you take enough matter and squeeze it small enough, you will eventually get so much gravitational force that nothing can prevent the matter from continuing to collapse.

In Einsteinian terms, the space around the matter becomes curved into a closed figure, but the result is the same: the matter is squeezed to infinite density. Long before it reaches that state, though, there is a region around the matter at which the escape velocity is greater than the speed of light.

The effect of that should be pretty obvious. If light can't escape, you can't see down into the hole. Moreover, anything that goes down in the hole can never come out: that is, if you accept the speed of light as the top limiting velocity of the universe, nothing can come out.

The area at which space is curved into a closed figure—or the region at which the escape velocity is equal to the speed of light—is known as an *event horizon*, and interestingly enough both Newtonian and Einsteinian equations give the same location to it.

It is the region at which $R = \dfrac{2GM}{c^2}$ (Equation One)

where R is the radius from the center, G is the universal constant of gravitation, and c is the speed of light. For our sun, that radius is on the order of 2 kilometers: if the sun is

160

ever squeezed that small, we'll never be able to see it again.

An observer diving into the black hole would never know when he had crossed the event horizon. He could continue to send signals to his friends outside, and as far as he could tell, they would go right on up and out.

Those outside the hole, though, can never under any circumstances receive information from inside it.

Now, as it happens, if we measure the total amount of matter in the universe, and plug that in for M in equation one; and we take the furthest object we can observe and plug that in for R; then the equation almost balances.

Almost, but not quite. There isn't enough matter in the universe; we're missing from 20 to 90%, depending on whose figures you use for M and R.

If the equation were to balance, space would be curved into a closed figure at the boundaries of the universe, and we'd live in a closed universe.

Eventually, in a closed universe, those galaxies receding from us will stop and come back, and the whole universe will be packed into a big wad at the center. What happens after that is debatable, but a number of cosmologists want badly to believe in a closed universe.

It also means, of course, that we live inside a black hole ourselves: that is, our whole universe is a black hole.

If we don't live in a closed universe, the receding galaxies will go right on receding, and this disturbs some theorists. Thus, Weber's coincidences were welcome in many cosmological circles. Others tried to build gravitational antennae to confirm his results.

Then a second startling result came out of Weber's shop. It appeared that there was a 12 hour sidereal cycle to the coincidences, and furthermore, that this cycle was related to the galactic plane. In other words, gravitational waves originated in the galactic center.

We have a good estimate of the distance to the galactic center, and thus were able to estimate how large an effect at the center of the galaxy would be required to deliver that much force to us out here on our spiral arm. The result was once again dismaying. Far too much energy was apparently being turned into gravitational waves.

Now the energy radiating from the galactic center could be either sprayed out in all directions, obeying the inverse square laws, or it could be "beamed" into the galactic plane. Obviously less total energy is involved if it is "beamed," but what mechanism might account for that?

The speculations were many, imaginative, and varied; they were also rather frightening.

* * *

Let's take a moment to go back to black holes. When matter gets dense enough to satisfy equation one, and the event horizon forms, things don't just stop there. The matter goes on collapsing; we just can't see it any longer.

In fact, *nothing* can stop the collapse. In theory, the matter should quite literally become infinite in density. Infinity is a troublesome concept: how can infinite density be present in a finite universe? The answer is obvious: in some respects, the matter no longer remains in the universe at all.

When gravitational forces have got to this point, we have what is known as a "singularity"; a point at which normal laws simply do not apply.

Actually, things are worse than that. Not only don't normal laws apply, but the relativity equations suggest that *no* laws apply. Strange things happen in the region of a singularity. Time is reversed. Conservation laws don't work. Causality is a joke: if you could get into the region of a singularity, you really could go back in time and assassinate your grandfather.

In fact, anything could happen, and science ceases to exist; and you don't even have to physically go to the singularity for this to take place. All you have to do is be able to observe one directly, and science has just gone down the drain. That bothers a lot of theorists and scientists, and rather disturbs me as well.

If there is a naked singularity—that is, a singularity not covered with an event horizon—then, at least in potential, there is no order to the universe.

Out of that thing might come ghosties and ghoulies and things that go bump in the night.

What, then, may we do to save science? Why, invoke censorship, of course.

The kind of censorship invoked is called rather whimsically the "Law of Cosmic Censorship," which states that "There shall be no such thing as a naked singularity." All singularities must be decently clothed with an event horizon.

Given cosmic censorship, a number of interesting laws about black holes may be proved: that they never get smaller, that if one is rotating it can't be sped up until the escape velocity is smaller than the speed of light, and a number of other rules that are collectively known as the laws of black hole dynamics.

Unfortunately, cosmic censorship deprives science fiction writers of some of their best stories.

It does it this way. If all black holes are covered with event horizons, it follows that you can't plunge into a black hole and come out elsewhere or elsewhen. Actually, if you plunge into a random black hole, all that could ever come out anywhere would be a stream of undifferentiated subnuclear particles; for all their fantastic properties, singularities do retain one feature, namely that gravitation in their region is rather high, sufficient to disassociate not only the molecular, but the atomic, structure of anything visiting them.

On the other hand, if a star about to collapse into black hole status is rotating fast enough, some solutions to the Einstein tensor suggest that the singularity formed will be a donut; you could dive through that and come out in one piece, provided the donut were large enough.

Large enough means galactic sized, I'm afraid; stellar size black holes will still get you too close to the singularity so that you can't use them for transportation. Furthermore, what you come out to on the other side is not, according to the equations, our universe at all. What it will be like, no one can say, except that it will have in it a copy of the black hole you dove through to get there.

So, turn around and dive back, of course; but that doesn't work. You go through and out again, all right, but into a third universe different from either of the other two. The black hole is still there, so try again—and come out in a fourth, and there behind you is that rather tiresome black hole again.

Is any of this real, or are we playing with ideas? No one really knows, of course. The most we can say is that the people who can solve Einstein tensors come up with that kind of result.

It's rather discouraging for science fiction writers. Here we thought we had a new way to get faster than light travel, what with black holes connecting us to another universe, or, just possibly, to another region of our own, and the very people who gave us the black holes go on to prove we can't use them.

Still, maybe there's a way out. Perhaps someone will find a solution. But they can't so long as the law of cosmic censorship is enforced, because singularities decently covered with event horizons can't come out and affect our universe.

Back to Weber and gravitation waves. One of the models constructed to account for the enormous gravitational energy generated in the center of the galaxy had a very large singularity lurking down there. Suns fell into it, and as they were eaten, gravity waves poured out. It was a rather depressing picture, our galaxy being eaten alive like that.

Then a number of other laboratories constructed gravitational antennae. Bell Laboratories, an English group, the

Russians, all made gravity wave detectors. In each case their equipment was supposed to be an improvement on Weber's.

None of them found any coincidences at all. People began to wonder just what Weber had done, and to doubt his results.

Last summer, at a Cambridge Conference of experimental relativists, the picture changed again.

The people who had built "improved" gravity wave antennae reported no results whatever.

Weber continued to report results, but with a change I'll get back to in a moment.

And two other groups, one at Frascati, Italy, the other at Munich, Germany, had built carbon copies of Weber's antenna. They got coincidences. Whatever Weber was observing, others have independently observed something similar now.

Meanwhile, Weber did a re-analysis of his coincidences, using a computer rather than human judgment to define just what was a coincidence. The result was startling. He still gets events—but they are no longer concentrated in the galactic plane. The sidereal coincidences have gone away, and with them has gone the evidence for the large singularity eating the galaxy.

Moreover, Dr. Robert Forward, of Hughes Research at Malibu, California, has constructed his own gravity wave antenna. Since lasers were invented at Hughes Labs, it's no surprise that Forward's antenna employs them. He has three big weights at the apexes of a right-angle triangle.

Lasers measure the precise distance of each weight from the others. A gravity wave will presumably distort that triangle, and thus be detected.

Forward has "events" too. They seem to coincide with the kinds of things Weber gets, but no serious attempt to compare results has been made as I write this.

For that matter, the Munich people have just got started. They were quite surprised, by the way; they'd thought

Weber's results were some kind of artifact.

It appears, then, that some kind of gravity waves do travel about through the universe; at least something that can affect large aluminum cylinders hundreds of kilometers apart is operating here.

The next step is to see if these events have any relationship to the bursts of x-ray energy detected by Vela satellites. At the moment that's not possible, and of course there are a lot more gravity wave events than x-ray events; but if the x-ray events are accompanied by coincidences on the gravity antenna, we'll know a lot more about both.

We may then be able to decide what gravity is: a force, or a distortion of geometry. We may be able to learn more about black holes, and what happens inside them, and who knows, those trips to alternate universes could be a real possibility.

Until we get rid of cosmic censorship, though, we'll never know what happens to the volunteers who go exploring down black holes.

Fuzzy Black Holes Have No Hair

Black Holes have no hair, but they're fuzzy. Because they're fuzzy, they're not really black.

If that seems confusing, read on: it doesn't get a *lot* clearer, but don't worry about it; not many people in all this world understand, and those that do have to believe three impossible things before breakfast.

In the previous chapter I told you that if a chunk of matter gets squeezed small enough it becomes a black hole. That's about 3 kilometers radius for our Sun. We also gave the equation for the radius for those who know how to punch the buttons on a good scientific calculator.

Actually we don't know the radius of a black hole. What we calculate is the radius of the "Schwarzschild region": that more-or-less sphere within which the surface gravity is so large that light cannot get out, and from which no signals can ever escape. In fact, *nothing* can escape from within the Schwarzschild radius. Also, down in there somewhere lurks a singularity that does strange things to time and causality.

* * *

If you watched a star collapse you'd never see it get quite so small as that. Instead it would appear that the collapse had slowed down and everything was now hovering just outside the Schwarzschild radius. The light would get redder and redder, and also dimmer and dimmer, and in milliseconds it would go out.

Since nothing can come out of the hole, we can't see in. We call the region from which nothing can escape the "event horizon," and we'll never know what happens inside because of the law of cosmic censorship, which was described in the last chapter.

Classical black hole theory dictates several laws of black hole dynamics. Some aren't too interesting, but the Second Law says the area of the event horizon can never decrease, and increases as matter and energy are pumped into the hole. This means that black holes never get smaller. Feed them matter and/or energy and they grow.

That lets us deduce one thing instantly. What happens if a normal matter and an anti-matter black hole collide? Well, nothing that wouldn't happen if two normal matter holes, or two anti-matter holes, collided, of course. The holes eat each other to form one larger than either, but we'll never know which ones contain normal or anti-matter. In fact, the question is meaningless.

You see, black holes have no hair.

This is a convenient way to say that everything we'll ever know about a black hole can be deduced from three parameters. Once you specify the mass M, the angular momentum J, and the electric charge Q, you've said it all. Nothing remains but location, which isn't important for the physics of the hole, but may be for the physicist who wants to study it.

Mass we understand. It doesn't really matter whether that mass is in the form of energy or matter; Einstein's

$E = mc^2$ takes care of that, and down in the hole it's irrelevant whether the rest mass is e or m.

Angular momentum comes from rotation of the object before it collapsed. Naturally it's conserved, so that if the star were rotating, the thing inside the hole rotates as well. It also rotates *fast*, just as a skater speeds up in a spin when she pulls her arms in.

The last parameter, charge, is just what it says, and it gives us a way to move a black hole around. If it isn't charged, feed it charged particles until it is, then use magnets to tow it.

The laws of black hole dynamics say you can never recover the rest mass energy (that's the Mc^2 energy, of course) of the original body. It's lost forever. Even shoving anti-matter down the hole gains you nothing.

However, you can get energy out of a spinning black hole. Up to 29% of the rotational energy is available, and in the case of a star that's a *lot*. To get it you throw something down the hole, and one of the things that comes out is gravity waves.

In our experience gravity waves are puny things, but we're a long way from their source. Up close is another matter entirely. You could be torn apart by them, as the characters in my story, "He Fell Into a Dark Hole," very nearly were.

Most of what we know about black holes comes from Stephen Hawking of the University of Cambridge. Many physicists think Hawking is to Einstein what Einstein was to Newton, and he's still a young man. This year Hawking has added quantum mechanics to classical black hole theory, and he's ruined a lot of good science fiction stories.

* * *

Somewhat over a year ago Larry Niven and I went out to Hughes Research Laboratories in Malibu. As stated previously the laser was invented at Hughes, so they do a lot of laser research there. They're also among the top people in ion drive engines, and they've done a lot with advanced communications concepts.

All that was fascinating, but we went to talk with Dr. Robert Forward, who's one of *the* experts on gravitation. I'd met him because he liked my black hole story ("He Fell Into a Dark Hole," nominated for a Hugo, but alas ...)

Bob Forward is the inventor of the Forward Mass Detector, a widget that can track a tank miles away by mass alone. (It can't distinguish between a tank at a mile and a fly on the end of the instrument, but if you use two and triangulate you're safe enough.) His detector can also be lowered into oil wells, or towed behind an airplane to map mass concentrations below.

After lunch we talked about black holes. Dr. Forward was particularly interested in Stephen Hawking's then new notion that tiny black holes might have been formed during the Big Bang of Creation. Since the Second Law predicts that they never get smaller, there should be holes of all sizes left. Some might be in our solar system.

They would come to rest in the interior of large masses. There might be quite a large one inside the Sun, for example, and even in the Earth and Moon as well. A very large mass hole, say 10^8 kilograms, would still be very small: about 10^{-19} centimeters radius. An atomic radius is around 10^{-9} cm., very large compared to such a hole, so that the hole couldn't eat many atoms a day, and wouldn't grow fast.

Black holes inside the Earth or Sun aren't too useful because they're hard to get at. Bob Forward wanted to go to the asteroids. You search for a rock that weighs far too much for its size. Push the rock aside and there in the orbit where the asteroid used to be you'll find a little black hole.

You could do a lot with such a hole. For example, you could wiggle it with magnetic fields to produce gravity waves at precise frequencies. There might be all sizes of holes, even down to a kilogram or two.

It sounded marvelous. Larry and I figured there were a

dozen stories there. I'd already written my black hole story, and Larry hadn't, so he beat me into print with a thing called "The Hole Man." All I got from the trip was a couple articles and columns.

Well, Larry's story (which won a Hugo. Sigh.) has just been reprinted in his new collection, while the columns I did about little black holes have been forgotten—I hope!

I'm glad I have nothing in print about tiny black holes, because Hawking has just proved they can't exist. Oh, they can be formed all right, but they won't be around very long. It seems that black holes aren't really black. They radiate, and left to themselves they get smaller all the time. The Second Law needs modifying.

Hawking points out that Eistein's general relativity, which produces most of the primary equations for black holes, is a classical theory. It doesn't take quantum effects into account.

Hawking corrects this. In quantum theory a length, L, is not fixed. It has an uncertainty or fluctuation on the order of L_0/L, where L_0 is the Planck length 10^{-33} cm.

Since there is uncertainty in the length scale, it follows that the event horizon of the black hole isn't actually fixed. It fluctuates through the uncertainty region.

In fact, the black hole is FUZZY, and energy and radiation can tunnel out of the hole to escape forever. It's the same kind of effect as observed in tunnel diodes, where particles appear on the other side of a potential barrier.

Since black holes have no hair, although they do have fuzz, the quantum radiation temperature—that is, the rate at which they radiate—must depend entirely on mass, angular momentum, and charge.

It does, but I'm not going to prove it for you. Hawking uses math that I *can* tool up to follow, but I'm not really keen on Hermetian scalar fields, and I doubt many readers are either. If you want his proof, send a dollar to Gravity Research Foundation, 58 Middle Street, Gloucester, Mass. 01930 and request a copy of Hawking's paper "Black Holes

Aren't Black."

Hawking shows that the temperature of a black hole is,
$$T = 10^{26}M$$
where M is mass in grams, and T is degrees Kelvin, and the lifetime of a black hole in seconds is

$$t_L = 10^{-26} \, M^3$$

Using my new Texas Instruments SR-50 that handles scientific notation and takes powers and roots in milliseconds, it wasn't hard to work up Figure 11 from these equations.*

There are more numbers than we need, of course. It's a consequence of the pocket computer. Not long ago I'd have had to use logs and slide rule, and I'd have done no more than I needed. Now look.

The first thing to see is that small holes have uninteresting lifetimes. In order for one to be around long enough to use it, the hole must be massive.

Any black holes formed in the Big Bang would be 10^{10} years old now; so if they weren't larger than a small asteroid they're gone already. Worse, that exponential decay rate defeats us even if we find a hole just decayed to an interesting size. It will still vanish too fast to use.

So there went Larry's "Hole Man", and two stories I had plotted but hadn't written, and I suspect a lot of other science fiction as well. Sometimes I feel a bit like Alice when she protested, "Things *flow* here so!"

But it's what we get for living in interesting times, and it ought to teach my friend Larry not to rush into print ahead of me...

---

*My SR-50 has long been shelved, of course. Now I use a full computer, and anyone can afford a better calculator than the SR-50 was. Marvelous times we live in ...

Figure 11

MASS, RADIUS, AND LIFETIMES OF BLACK HOLES

Description	Mass (grams)	Radius (cm)	Lifetime (seconds)	Lifetime (years)
Kilogram	1000	1.48×10^{-25}	10^{-19}	----
Billion gm.	10^9	1.48×10^{-19}	.1	----
2365 tons	2.15×10^9	3.19×10^{-18}	1	1
---	6.7×10^{11}	9.9×10^{-17}	31.5 million	million
---	6.7×10^{13}	9.9×10^{-16}		million
---	1.5×10^{15}	2.18×10^{-13}		10^{10}
Ceres	8×10^{23}	1.2×10^{-4}		10^{38}
Earth	6×10^{27}	9×10^{-1}		Eternal
Sun	1.99×10^{33}	3 kilometers		
Galaxy	10^{11} suns	3×10^{16} = .03 lightyear.		
Universe	10^{22} suns	2.95×10^{27} = 3 billion lightyears.		

Crashing Neutron Stars, Mini Black Holes, and Spacedrives

The universe is a queer place. I know that doesn't surprise most of the readers of this book, so let's look at some far-out theories before the relentless march of science catches up and converts them into (yawn) just more engineering.

There's also a point to be made. At least once a week I get a new "theory" from one of my readers. Sometimes it's blueprints for a device that gives out more energy than it takes in, sometimes it's a refutation of Einstein, sometimes something else, but in nearly every case there's a common factor: the cover letter says no one will *listen*, and I'm the inventor's last resort. Generally, too, I'm asked "Why won't they listen?" but the question is rhetorical; the next paragraph tells me that orthodox scientists do not have open minds, are afraid of new ideas, etc., etc.

There's a little truth to that, but not so much as is often supposed. Orthodox science can get pretty far out.

Everyone knows what a neutron star is, right? A sort of intermediary superheavy object: that is, squeeze atoms hard enough but not too hard and they'll form a superdense goop. Everything has been shoved together, but it's still ordinary matter. That's what dwarf stars are made of.

Squeeze more and the atoms can't stand it. The electrons are forced out of their orbits and down into the nucleus. When you push an electron hard into a proton you get a neutron, and a sufficiently squeezed object will be nothing *but* neutrons, i.e., a neutron star.

Finally we can continue squeezing (the easy way is to pump more and more matter onto a neutron star and let gravitational collapse do the work) and the very neutrons can't stand it: they're squeezed right out of the universe. Well, maybe not: but the result is a black hole, an object whose surface gravity is greater than the speed of light so that it can never be observed, and that's a very queer thing indeed. For further details consult some of my previous chapters, or there are any number of recent books.

Surely those theories are far out? And for a very long time they were pure theory. Back in the 30's J. Robert Oppenheimer deduced the possibility of neutron stars from an analysis of Einstein's work on gravitation. He also noted (as had others) the possibility of black holes. He published these speculations. No one laughed. He wasn't thrown out of the union. Why? Why did "they" listen to him, but "they" won't pay attention to outsiders?

Well, of course there is a certain degree of old-boy networking here; obviously you pay more attention to people you know or have heard of than people you don't know and who don't seem to have any qualifications; but that's not the whole of it by a long shot. Neutron stars and black holes were far out and queer indeed—*but they did not challenge basic physics*. Quite the contrary: they were deduced from accepted ideas.

Meanwhile, back in the observatories, the radio

astronomers came up with a queer result: pulsars. These were very small objects which emitted quite a lot of energy at fantastically regular intervals on the order of a second. They were entirely unexpected: no theory predicted them, no theory accounted for them, yet they had to be explained. And of course explained they were, because it's easy to show that neutron stars will rotate very rapidly, they can send out bursts of radio energy, and even better, they'll be slowing down (very slightly) all the time. Back to the observatory to discover that the pulsars were slowing down at just the right rate (micro-seconds a year) and lo, Oppenheimer's far-out theory becomes universally acceptable.

That took care of neutron stars: and if *those* probably existed, then why not black holes? Thus the sudden interest in holes.

Now there are several ways we can create black holes. The simplest is for a properly-sized star to "go out": that is, the star runs out of fuel and begins to collapse from simple gravitational attraction. Of course the collapsing process will itself produce heat for quite a long time, and that brings us to another far-out theory: has the Sun, *our* Sun, "gone out"?

It may have. If it had we wouldn't notice, because it would only have to shrink a few kilometers a year to give off the energy it does. (Gravitational potential energy is *powerful* stuff.) Moreover, the continued collapse would eventually cause it to re-ignite, halting the collapse. There are a very few theorists who seriously propose this alternation between collapse and hydrogen burning as an explanation for Earth's Ice Ages. The theory is not widely held, but those who propose it aren't laughed at. Why not?

Well, first, the Ice Ages are real, and we don't have a convincing mechanism to explain them. Second, out in the old Homestake Mine, deep underground, they're searching for solar neutrinos—and they can't find them. If the Sun is truly burning hydrogen, there ought to be a lot of neutrinos, and there's no good reason to doubt that the Home-stake apparatus would trap them—so where are they? The "Sun's gone out" theory does explain an experimental result.

Leaving that particular theory let's get back to black holes. There are other ways they might be formed. One would be to focus a laser small enough. After all, E equals mc-squared works both ways: if matter can become energy, energy can become matter, and thus a sufficient energy density would make a black hole. There'd be no way to tell it from other holes, either. This is not a very practical way to get black holes.

Finally, we can make black holes by having really humongous pressures. Stephen Hawking of Cambridge first postulated the idea of mini-black-holes, little tiny things massing from a few to a few million kilograms; previous to Hawking's work nobody contemplated a black hole smaller than stellar size.

Hawking thought that mini black holes had been formed during the Big Bang, when there were certainly sufficient pressures for creating them. Since it was then thought that black holes never go away once formed, we could expect in a few decades to go looking for primordial black holes in the asteroid belt.

Having created mini-holes, Hawking then proceeded to destroy them. He applied quantum mechanics to black holes and demonstrated that the things aren't stable. Black holes evaporate, small ones doing so rather quickly and quite violently. (See the previous chapter. I'm particularly proud of that one because it was the first popular-press publication of Hawking's evaporation-theory; now his view is universally accepted.)

The upshot was that holes massing less than 10^{15} grams when formed during the Big Bang are all gone now, and any larger holes formed then (and now getting down to small size) will vanish spectacularly with much high-temperature radiation. Consequently, we needn't bother looking for small black holes.

Well, guess what? Mini-holes are back again. Kenneth Jacobs and Patrick Seitzer have just published a prize-winning essay submitted to the Gravity Research Foundation (for a copy of their paper send a couple of dollars to the GRF at 58 Middle St., Gloucester, Mass. 01930) entitled "Mini Black Holes Are Forming Now." According to Jacobs and Seitzer the little holes are generated inside neutron stars.

It works this way. Take one common or garden variety neutron star. Pile matter on it. You can do that by putting the star in a dusty region of the galaxy, or by giving it a normal-matter companion in close orbit: the companion loses mass to the neutron star.

As the neutron star grows, a "pressure and density spike" develops in the interior. If you keep piling on enough matter the neutron star will simply collapse into a large black hole, but let's consider a time just prior to that: a time when the interior of the neutron star is far denser than its surface.

Jacobs and Seitzer offer mathematical proof that it's at least possible that the interior collapses into a mini-hole, leaving the rest of the neutron star intact!

Once the mini-hole is formed, it begins to evaporate. On the other hand, it's inside a neutron star, surrounded by all that lovely dense matter, and it can begin to feed. By eating neutrons the hole grows larger. Thus we have two counteracting tendencies: the hole eats the star, growing larger, and also undergoes Hawking "evaporation." Under certain conditions the two may just balance each other. It's even possible that the tiny hole will prevent the formation of a big black hole, thus giving the neutron star a few billion years more of normal life.

Sigh. It's hard to see how those holes can be useful. There are all kinds of marvelous things you can do with a mini-hole if you can get at one, but the interior of a neutron star is a well-protected place; nature's safe-deposit box.

Back in the old days before Hawking evaporation, we thought mini-holes might come to rest inside asteroids, and it would be no great trick to go move the asteroid to get at the hole; but doing that to a neutron star would be a bit more difficult.

It would also be dangerous. The hole exists in equilibrium, eating the star at a rate to balance its evaporation. Disturb that and the entire neutron star could become a black hole in milliseconds! It would also toss out a bit of energy: 10^{55} ergs. (For comparison, our Sun puts out 10^{39} ergs each year, and an ordinary nova gives off 10^{44} ergs.)

In fact, Jacobs and Seitzer describe ways these mini-holes might be the driving mechanism for super-novae. They also speculate that this could be a mechanism to explain gamma-ray bursts, and here we leave the realm of theory to return to the real universe. Gamma-ray bursts exist, and their spectra are consistent with events predicted by the Jacobs and Seitzer mini-hole theory.

Now *that* gets interesting. As MIT's Phillip Morrison put it recently, x-ray astronomy strives for predictive power, but so far is more like social science than physics: we can find ways to *explain* observations, but not *predict* them. Of course Jacobs and Seitzer haven't made a prediction either, since they knew of the x-ray bursts before they devised their mini-hole theory; still in all, they've as good an explanation as anyone.

Those x-ray events are bothersome. They're also hard to study, because of the atmosphere. Of course that's fortunate: if we could directly observe those x-rays coming from space, we'd none of us be here. Our atmosphere lets ordinary light squeak through with 20% attenuation, but it stops x-rays cold, and that's just as well for us, but bad for astronomers. Thus it's only since we began sending up satellites that we learned of x-ray events, and we still don't know much about them.

What we do know is driving astrophysicists nuts. Astronomy, after all, is usually concerned with rather slow and stately phenomena, and highly repetitive events, such as orbits. Things aren't supposed to change very fast. Now suddenly we're confronted with rapid x-ray events, and to make it worse we can't focus the x-ray satellite "telescopes" very well, so we can't even be sure of where the x-ray bursts are coming from. However, the Dutch astronomical satellite has a rather small field, and happens to point in a good direction, and it says that a probable source of x-ray bursts is globular clusters. And that's weird.

Globular clusters: big balls of very old stars. About 200,000 stars to the ball, packed very close with average interstellar distances of light-months. There are a number of such clusters hanging about the galaxy in a kind of halo. Being *very* old stars they ought to be stable, without too many strange things happening; but here are these x-ray bursts.

The bursts themselves look like a spike of energy, then a slow decay. Typical: every 4½ hours the x-ray energy goes from essentially none to a lot in about 0.1 second, the event lasts at peak for about 10 seconds, and then decays over a period of hundreds of seconds. That needs explanation.

Possibly the Jacobs-Seitzer theory will cover the facts. There are other theories, very far-out, *weird* in fact, probably wrong, but really beautiful. Take, for example, crashing neutron stars.

Imagine a rather floppy disk of gaseous matter orbiting a star like the rings of Saturn. Put this whole mess into a globular cluster. The disk probably isn't that hard to come by in there: globular clusters act as gravitational traps, and might easily accumulate all kinds of cosmic junk, including the debris of old stars, etc.

Now image a neutron star in an orbit tilted with respect to the plane of the floppy disk. Every half-orbit the neutron star crashes through the disk, producing x-ray bursts. A *lovely* idea.

Alas, it's almost certainly wrong: there are many sources

of x-ray bursts, and it's just hard to believe that anything as unlikely as crashing neutron stars are *common*, even in our queer universe.

Let's try another theory: flash-burning. Take one neutron star. Let hydrogen fall on it. The hydrogen accumulates on the surface, slowly building up until the neutron star has an "atmosphere" a few millimeters thick. The density increases sharply until it hits the critical point, and bang! Every few hours there's a hydrogen bomb enveloping the neutron star.

Far out, right? Yet those theories were described at a meeting of the super-orthodox AAAS. Now let's look at one "neglected" theory.

Dean Drive: science fiction readers must know about that. Norman Dean was a crackpot inventor. Many years ago he built a gadget. (He also took out a patent; on that, more later.) Dean claimed something that sounds very reasonable, hardly far out at all: that his mechanism "converts rotary acceleration into linear acceleration." After all, a rotating object does have acceleration, right? Acceleration is acceleration, right? Thus to convert from one to the other violates no conservation laws.

Now that's an attractive concept. Take an electric motor (Dean used an ordinary quarter-inch electric drill) as the source of rotary acceleration; hook up to the Dean mechanism; and lo, you have a spacedrive. Very attractive, because by installing a Dean Machine in a nuclear submarine we'd have a spaceship already built!

Now Dean did build a machine. It did *not* lift itself off the floor. Observers agree on that much—and that's about all they do agree on. The machine shook and shimmied and jumped up and down, and there was a famous photograph of Dave Garroway shoving a piece of paper under the machine. Also, John W. Campbell Jr. reported that he'd seen the apparent weight of the Dean Machine, as measured on a bathroom scale, appreciably decrease when the gizmo was turned on, to be restored when it was switched off. G. Harry Stine reports that he felt the machine push against his hand; push hard, and when it was turned off the push wasn't there.

Unfortunately, that's about the sum total of observational evidence. Dean never let his machine be examined by anyone else. However, the story doesn't quite end there.

Not long ago Harry Stine published an article about the Dean Drive in that other magazine (the one with rivets). Robert Prehoda (DESIGNING THE FUTURE; Chilton, 1967, among other excellent books) and I were discussing Harry's article (Harry is an old friend of Prehoda's) and it came out that Bob Prehoda had tried to buy the Dean Drive back in the early 60's. At the time Bob was representing the Rockefeller family, so the ability to pay real folding money wasn't in question. I knew of a couple of other aerospace firms who'd also made the effort. The stories were remarkably similar: Dean wouldn't let anyone examine the machine. He wanted a million dollars and a Nobel prize up front: *then* you could play with the gadget. No one was going to put up that kind of money without seeing the gizmo in operation. There are just too many ways the reported results could be obtained without anything new. For example, if you jump up and down on your bathroom scale at just the right frequencies, you can fool the scale into thinking you're either lighter or heavier than you are; its response time just can't handle that non-steady weight.

Prehoda had also known Dr. William Davis, USAF colonel and successful inventor, who had worked out a physics theory which, supposedly, allows the Dean Drive and other "reactionless" drives to work. "Spacedrive" Davis spent a lot of time promoting what came to be called "Davis mechanics," and although his practical work made him a good bit of money, he acquired an unenviable reputation among theoreticians. Harry Stine worked for Davis and built several gadgets supposed to test the Davis theories.

Davis is dead. Norman Dean is dead. John W. Campbell

is dead. And neither Prehoda nor I could get out of the back of our minds a simple fact: Gregor Mendel discovered genetics but the results lay unused from 1868 to 1900 because no one wanted to listen to this crazy abbot. Could something similar be happening here?

We doubted it, but it did seem reasonable to try to get all the data together in one place before everybody who actually saw the Dean Machine, or tested Davis's theories, went off to Murphy's Hall. We decided to invite Harry to Los Angeles to confer with some other people on how best to test this whole concept once and for all. Understand: none of us, not one, really "believes" in spacedrives; but the concept is so blasted important and the consequences are so far-reaching, that surely it's worth a little effort?

So who do you invite to such a conference? What's needed are people with minds that are "open," but not minds pierced with gaping holes; with enough knowledge of physics and math to follow the arguments; and enough scientific integrity neither to bite on the idea simply *because* it's unorthodox nor to reject for that reason.

We ended up with: Robert Prehoda, chemist and propulsion systems expert; Larry Niven; Dan Alderson, astronomer and computer scientist; Robert Forward, physicist and gravitation expert; G. Harry Stine, engineer and gadgeteer extraordinary; and myself. It was an interesting lunch. Something may come of it. And the result illustrates precisely the point of this chapter.

The general reaction was simple. Davis's theoretical papers are interesting, but not terribly valuable. They assert that there's something wrong with physics. Okay, and maybe there is; but contemporary physical theory has a lot of scalps hanging from its belt. Maybe it needs changing, but not without some convincing evidence; and mathematical theories *are not evidence*.

That's what I've found very hard to get across to people. In order for a "Dean Drive" or the gizmos postulated by "Davis mechanics" to work we really do have to throw out a very great deal of very fundamental physical theory. You *cannot* simply "convert rotary acceleration into linear thrust" and remain consistent with what we think we know. Spacedrives are just *impossible* given current theory. If you want to move a ship in inertial space, you have got to throw reaction mass out the back end, and that's all there is to it.

Now, sure, theory can be wrong. Einstein did his thing, and Newtonian physics has never been the same. We've yet to come to the end of the changes wrought by quantum mechanics. Moreover, what with 100 and more elementary particles kicking around, many believe that physics is *due* for a restructuring along the order of the changes rung in by Einstein.

But note: Einstein didn't generate his theory out of pure math. Far from it. His contributions came in explanation of *observed* phenomena that simply couldn't be explained by Newtonian concepts. The change in the perihelion of Mercury; the photo-electric effect; the Michaelson-Morely experiment; all *these* said, loudly, that there was something wrong with physics and it was time to get up a new theory.

Davis, on the other hand, played about with math and came up with "predictions" absolutely impossible within present physical theory. Is there any wonder that no one takes them seriously? Now a working "Dean Machine" would change all that. If you can actually build a gizmo that produces thrust without throwing mass overboard— even a *tiny* thrust—then there's nothing for it: physics is in trouble, and new theory *must* be found. Without such evidence, though, there's simply no reason to revise physics theory. After all, the "orthodox" stuff works quite well. It fits the observed universe.

Now: there remains Harry Stine's memory, after all these years, of that machine pushing against his hand. One result of our conference was a consensus that if the Dean Machine did not work, it at least did not employ any of the common means of producing spurious results. It didn't have feet that periodically touched the floor, and such like. That doesn't mean it worked the way Dean said; it just means we don't know how the result was obtained.

Then too, when Harry Stine was working for William Davis, Stine built a couple of gadgets (see Harry's *Analog* article for details) that gave results you certainly wouldn't have predicted in advance. They *may* be explainable in "normal" theory, but they are a bit queer, and they do fit Dr. Davis's theories.

The upshot of our meeting was this: the theoretical stuff can wait. What's important is that Harry Stine, as one of the few men now alive who actually saw the Dean Machine and also worked with experimental gadgetry to test Davis mechanics, get his *experimental results* into proper form and publish them. *That's* what's important. *Evidence.* As Bob Forward observed, once the plans are published, one of these days a research physicist will build the apparatus: and if several different labs get the results Harry did, and have no conventional explanation for them, *then* it will be time to trot out the theories. "One good experimental result is worth a thousand theories," Forward said; and he's right.

Back in the old days a number of us built Dean Machines from the patent specifications. We had technicians and they had to be paid between jobs; might as well use them to test new ideas. None of the gizmos worked. I now know of four that were built. Harry Stine can explain that: Dean, being, uh, highly suspicious, didn't describe his actual gadget in the patent. Of course that's stupid, because a patent protects only what's disclosed, but it's very much in keeping with what's known about Dean's personality.

Harry says the actual Dean Machine was the goldarndest collection of springs and slipclutches and mechanical linkages he's ever seen in his life. He also says it pushed hard against his hand, and he'll never forget that. Harry thinks Dean had something. The question is, can it be reproduced? Did it "work" as Dean thought, or did it merely act strangely? There are ways to test that, unambiguously. Until that's done, though, theory is not relevant.

So it's that simple. It isn't that the "establishment" won't *listen*, as so many would-be theorists insist; it's that the newcomers insist that physicists *only* listen. They have nothing to *show*.

Now I get all kinds of blueprints and plans and equations from my readers. Those who send them assume, I hope rightfully, that (1) I know something of what science and technology is all about, and (2) I try to keep an open mind. I'm willing to listen.

But please, all of you who have new ideas, keep in mind what I said earlier. If you have plans for a perpetual motion machine and you really believe it will work—why build it! Don't send out plans only and then complain that "nobody listens." Of course nobody listens; it takes a lot of effort to spot the flaw in a very complex device (one I was sent ran to fifteen pages of drawings) but the chances are good that the flaw is there. I don't care how good a theory you have to prove that you can get energy out of your swimming pool; but I care a lot if you have built the device and it works.

Experimental results. Build the device. Make it work. And *then* if nobody listens, something can be done. Certainly "orthodox" physics doesn't know everything. I've said myself that I believe (emotional bias only; I have no evidence) that we'll someday build faster than light ships, and yes, I suspect there might even be "spacedrives."

But we won't find them from blueprints. It takes evidence. Once you've got that, you'll find plenty of people to listen.

"In the Beginning . . ."

First, let me establish something. When I go to Cal Tech I do *not* expect an experience out of H. P. Lovecraft. Horror may be interesting at the proper time and place, but it's not very pleasant as a total surprise.

It started peacefully enough. Dr. Robert Forward, the Hughes Research gravity expert you've heard of here and other places, called to ask if I would be interested in meeting Stephen Hawking. Since Hawking is thought by important physicists possibly to rank alongside Newton and Einstein, it took perhaps five milliseconds to think over the proposition. I didn't even need to look at my calendar; nothing I had planned could be that important.

A week later Larry Niven and I drove over to the California Institute of Technology. It was a bright spring afternoon . . .

In order properly to tell this story I must now give some personal details about Professor Hawking. I've consulted his friends, who assure me that he doesn't mind.

Stephen Hawking is quite young, early thirties at the oldest. He is a resident theoretician at Cambridge University, and he yearly produces marvels in astronomical theory, particularly in the field of black hole dynamics.

In "Fuzzy Black Holes Have No Hair" I described Hawking's marriage of quantum mechanics to Einstein's classical relativity theories to produce the startling prediction that black holes are unstable. He is also responsible in large part for the so-called laws of black hole dynamics. An important man indeed.

Alas, Professor Hawking suffers from a nervous-system disorder which severely impairs his speech and confines him to a wheel chair. Those who attend his lectures are warned that they must listen closely; he can be understood, but only with difficulty and concentration. Of course this would be true if he spoke with the oratorical clarity of William Jennings Bryan to such bards of the sciences as Larry Niven and me, so we were prepared to be doubly confused.

Cal Tech's architecture is a neat blend of Old California and modern LA; arched thick-walled Monterey-style buildings with large shaded porches alternate with steel-and-glass towers and clean-lined functionalism. It sounds horrible, but the effect is actually quite pleasing. It's a nice place to be, especially if you're looking forward to hearing one of the truly great men of our time.

The lecture was in a small modern slant-floored room of the type sometimes called lecture theaters; the sort of classroom lecturers like. The tiered seats let everyone have a good view of the speaker and his demonstration materials, and give the speaker a good view of the audience.

It was only partly filled: graduate students, several undergraduates, a sprinkling of faculty, one or two of the top names in theoretical physics. It was a room of serious women and men, mostly younger than I, all expectantly quiet. At the bottom of the well, the focus of attention on the stage, was an incredibly thin, very young-appearing man seated in a high-backed motorized chair of Victorian design; the chair had no flavor of the hospital about it. He wore a light suit, dark shirt, and flowered tie, and he kept his hands folded carefully in his lap as he was introduced.

The chairman gave his credits and spoke wonderingly of how privileged we were to hear a man of this stature. No one disagreed. Not, of course, that anyone would have said anything no matter what he thought, but the total silence in the room was an obvious sign of unanimous assent.

Hawking began to speak. Everyone leaned slightly forward, straining to hear. Except for the heavily slurred voice there was absolutely no sound; you could quite literally hear a pen drop, for I dropped mine and it clattered loudly on the cement floor.

This is the scene, then: a lecture room partly filled with very bright people, a few extremely well known in theoretical physics, others students at one of the world's most prestigious institutions. They all strain to hear a wizened young man who makes awkward gestures and speaks with a thick slur that keeps his words just at the edge of intelligibility.

He grins like a thief. He's obviously not in pain, and he doesn't feel sorry for himself. And he tells that room of bright people that everything they thought they knew is nonsense. And he chuckles.

He tells us that the pudding that ate Chicago may some day exist; that duplicates of each one of us may one day wander the universe; that *anything* can, and probably will, happen. He tells us that the universe isn't lawful, never will be lawful, never *can* be lawful; that we *cannot ever* know enough to predict the totality of events in this universe; that at best we study local phenomena that may be predictable for an unspecifiable time.

And he laughs.

He tells us that Cthulthu may exist after all.

As I said, it was an afternoon of Lovecraftian horror.

Larry and I escaped with our sanity, after first, in the question period, making certain that Hawking really did say what we thought he'd said.

He had.

* * *

Stephen Hawking's lecture had originally been entitled "The Breakdown of Physics in the Region of Space-Time Singularities." The title was flashed on the screen; then another slide took its place, and Hawking chuckled. The new slide said:

PHYSICISTS
THE BREAKDOWN OF ~~PHYSICS~~ IN THE
REGION OF SPACE-TIME SINGULARITIES

He began simply enough. The principle of equivalence, he said, is well established. This is the principle that states that *inertial* mass, that is, the resistance of objects to being moved by an outside force, is exactly equivalent to *gravitational* mass, that is, the gravitational force a given mass will exert. There are not two kinds of mass.

This was Galileo's principle, and there's the famous apocryphal story of his dropping a cannon-ball and a musket-ball from the Leaning Tower of Pisa and observing their striking the ground at the same time. Obviously if gravitational and inertial mass were different, heavy objects would *not* fall at the same speed as light ones.

So far so good. Next, gravity affects light. It can bend light rays, as predicted by Einstein and observed several times in solar eclipses.

Now in short order: the energy-momentum tensor of gravity is positive; gravity is universally attractive, not repellent. Therefore, enough mass will create a field from which no light can escape.

The Special Theory of Relativity says that nothing can travel faster than light.

And *therefore* sufficient mass must create a space-time singularity, a place which cannot be observed.

A singularity is therefore inevitable; that is, at least one singularity must exist, provided only: (1) that Einstein's general relativity is correct; (2) gravity is truly attractive and never repellent; and (3) enough mass has ever been collected together.

And *therefore* at least one singularity exists in our universe, since at the time of the Big Bang all the conditions certainly prevailed; and also, it's very likely that other singularities have been created by collapse of stars, since many stars have more than enough matter and don't have enough energy to throw that matter away as they die.

* * *

OKAY so far? Nothing startling here. Bit dry, but all we've shown is that singularities must exist, and nearly everyone accepts the idea now. They're hidden away inside black holes, of course, and observers are now very nearly certain that we can *observe* a black hole.

Well, not observe the hole itself; but Cygnus X-1, an x-ray emitting star in the constellation Cygnus, has an invisible companion and the pair of stars, the one we can see and the one we can't, together act very like what Cal Tech's Kip Thorne predicted such a pair would act like if one were a black hole.

So what else is new? We've proved black holes can exist, and lo, the observers think they've found one. What's scary about *that?*

Nothing, so far. Holes aren't scary unless you're about to fall into one. We even understand them. We know they "have no hair," that is, that they can be completely described given their mass, M; angular momentum, J; and electric charge, Q, Given these data we can describe their shape, and predict what effect they'll have on nearby objects, and play all kinds of fascinating scientific-theory games.

We can talk about black hole bombs, and toy with ideas

on how to extract energy from them: take one rotating black hole, throw garbage into it, and you not only get rid of the garbage, but can get useful energy back out. There are speculations (not SF; just plain science) about extremely advanced civilizations using black holes for precisely that purpose.

There's just no end to the nice things you could do with black holes, and although not many years ago they were no more than toys for theoreticians to play mental games with, black holes have become household-word objects now.

Black holes don't make us nervous.

Ah, but inside each black hole there lurks a singularity. This is the little beastie that breaks down physics in the nearby regions. By definition they do things we can't predict. They behave in strange ways. Up close to them time reversals can happen. How, then, can we avoid this breakdown of our nice predictable universe?

Hawking discussed several theoretical alternatives, and dismissed each. A couple of the cases seemed to startle one of the big-name theoreticians listening to the lecture. When Hawking was finished, though, the singularities were back and inevitable. I won't pretend to have understood all of this part of the lecture; and I wouldn't bore my readers with it if I had. If you appreciate that sort of thing you'll read Hawking's paper when it comes out.

For the rest of us I sum up by saying that he found no good alternatives; eliminating General Relativity doesn't eliminate the singularities, or else lands you in an even worse theoretical soup.

Therefore, let us look at General Relativity; but let us add quantum theory to it. Hawking recently published that work, and I described it here.

The important fact is that the quantum effects violate cosmic censorship. The Law of Cosmic Censorship, you may recall, states that there shall be no naked singularities; every singularity shall be decently clothed with an event horizon that prevents us from ever being able to observe it directly, and thus prevents us from observing the region in which physics breaks down.

Thus we needn't fear the singularity. It can't affect our lives, because nothing it does can get out of that black hole "around" it.

But adding quantum effects to General Relativity repeals cosmic censorship. Black holes evaporate. Big ones slowly, small ones rapidly, all inevitably. And what of the singularity that MUST have been created by the Big Bang of creation?

Evaporation of black holes produces naked singularities. We may play about with the concept of quantizing relativity, and Hawking did; but the conclusion was inescapable. Again I don't pretend to have followed every step, nor did most of the rest of us in that room; but several did, and they weren't pleased.

Because now comes the punchline. The singularities emit matter and energy. And "they emit all possible configurations with equal probability. Thus, perhaps, this is why the early universe from the Big Bang singularity was in thermal equilibrium and was very nearly homogeneous and isotropic. Thermal equilibrium would represent the largest number of configurations."

But since that time the universe has changed, and we have stars and planets and nematodes and comets and people; but the singularity must still be around. It emits. And what comes out is completely random, absolutely uncorrelated. This fundamental breakdown in prediction— Hawking is saying not only that we can't predict *now*, but that in principle we can *never* predict, no matter how much we know or how smart we get or how large a computer we build—is a "consequence of the fact that General Relativity allows fundamental changes in the topology of space-time; that is, allows holes.

"Matter and information can fall into these holes—or can come out. And what comes out is completely random and uncorrelated."

The hole can emit anything. Anything at all.

* * *

"No," I thought. I looked to Niven. "No," he was thinking. Surely we misunderstood.

And the thin chap grinned ever more broadly. "Of course we might have to wait quite a while for it to emit one of the people here this afternoon, or myself, but eventually it must—"

Hawking chuckled and waited expectantly, and after a long and very silent pause first one, then another joined him in laughter; but it had a rather hollow sound, or so I thought. Larry agreed when we could talk about it later.

So far as we can tell, we've just heard one of the top people in theoretical physics tell us that we don't know anything and can't know anything; that causality is a local phenomenon of purely temporary nature; that time travel is possible; that Cthulthu might emerge from a singularity, and indeed is as probable as, say, H. P. Lovecraft.

Hawking concluded by reminding us that Albert Einstein once said "God does not play dice with the universe."

"On the contrary," Hawking said, "it appears that not only does God play dice, but also that he sometimes throws the dice where they cannot be seen!"

* * *

Lovecraftian horror indeed. Our rational universe is crumbling. Western civilization assumes reason; that some things are *impossible, that's all*, and we can know that; that werewolves don't exist, and there never was, never could be, a god Poseidon, or an Oracle that spoke truly; that the universe is at least in principle discoverable by human reason, is *knowable*.

That, says one of the men we believe best understands this universe, is not true. It's not very probable that Cthulthu will emerge from the primeval singularity created in the Big Bang, or that Poseidon will suddenly appear on Mount Olympus, but neither is *impossible*; and for that matter, this world we think we understand, which seems to obey rational laws we can discover, isn't very probable either—isn't, in fact, in the long run any more probable than a world that includes Cthulhu, or the pudding that ate Chicago.

* * *

Well, of course I don't believe that; not in the sense that I'm going to alter my life to conform to a lawless and unpredictable universe. But I am now reduced to an act of faith: an irrational belief that the world and universe are, must be, lawful, and rational.

This is not "faith in science" or believing in science; not any more. It never was, actually; but Hawking has laid bare the hidden flaw. So long as science itself concluded that the universe was lawful, few of us were tempted to ask *why* this should be so, or to realize that this is the one question science can never answer.

Now, though, science itself says the universe is not lawful. If you want a lawful universe, you've got to take a leap of faith; you've got to hold fast to an irrational belief. While you're doing that, why not also believe there's a higher purpose to it all?

Is it harder to believe the universe is lawful and purposeful than to believe it is lawful but without purpose?

* * *

In the Beginning, the Big Bang emitted Chaos; and the Chaos was without form, and void, for it was homogeneous and isotropic. And the Singularity moved upon the face of the Chaos and emitted light; and the Universe was no longer homogeneous, for the light was divided from the darkness.

And there came forth firmaments and dry land and seas and stars and moons; and the worlds brought forth grass, the herb yielding seed, and the fruit trees yielding fruit after his kind, whose seed is in itself.

It is quite literally true that if you can believe that, you can believe anything; more, you *must* believe anything. To

195

exclude anything you must make an act of faith.

* * *

As we drove away from Pasadena, Larry remarked that if we ever had proximity to a singularity, he could well imagine people praying to it. After all, their prayers probably wouldn't influence what came out of it—but they might, and certainly nothing else would. I even had an idea for a flippant story to be entitled "The Oracle."

I don't think I'll write that story.

If this new work of Hawking's holds up—and if we've correctly interpreted what we heard—there are going to be some changes in the fundamentals of Western Civilization.

Will philosophy once again become the "Queen of Sciences"? I don't know; I suspect, though, that what we heard during our Lovecraftian afternoon will have a long reach. We either need some fundamental new breakthroughs in theoretical physics—and I've heard no hint of what they may be—or we'll have to start thinking about faith again.

Afterword to Part III

Although most of this was published several years ago, there has been little need for revision; surprising, given the progress of science.

At the 1978 meeting of the AAAS Dr. Joseph Weber reported that he still has events; he is reanalyzing his older data; and so far it is too early to determine whether the 12-hour periodicity really exists or was an artifact of the data analysis. Thus there's nothing new to report.

Since my spacedrive column was published I have corresponded with a consulting engineer who was invited to test a "spacedrive"; he wrote eagerly to tell me that hanging the fool thing on a pendulum and turning it on produced a very small, but very real, displacement from vertical—which is to say, there was a definite non-Newtonian force! That was exciting.

I was set to go up and view this marvel, but a few days ago I received a letter: my engineer correspondent had put a plastic bag over the apparatus; and the tiny "thrust" had vanished. Evidently there was some kind of standing wave effect. Sigh.

So that all the evidence on spacedrives is as described in the essay. I wish there were more, but there isn't.

As to Hawking's oracles: the last place I know of that he presented that paper was to a Vatican conference of Catholic scientists. I expect they found it as disconcerting as I did. And no one has refuted Hawking's equations.

A STEP FARTHER OUT Part 2

It is hard to over-praise this collection of articles on the latest developments in science, or their author. But I'm going to try. Of all the people I have ever met in my life, Jerry (Pournelle) is one of the most colossally educated in science. He has advanced training in systems engineering, psychology, physics, mathematics, logic and political science. He is the only science degree person I have ever known who was able to explain coherently the entire Velikovsky controversy, an hour later do the same for the Atlantean legend, and then, in a major talk, describe the new, dynamic, holographic view of the structure of the brain.

A. E. van Vogt

Jerry Pournelle worked on both the Mercury and Apollo space programmes, and was involved in the qualification tests for the original astronauts. He has degrees in engineering, psychology and political science, and is the winner of the 1972 John Campbell Award.

Also in *Star*:

A STEP FARTHER OUT

Part 2

by Jerry Pournelle, Ph.D.

Preface by Larry Niven
Foreword by A. E. van Vogt

Star

A STAR BOOK

published by

the Paperback Division of
W. H. ALLEN & Co. Ltd

A Star Book
Published in 1981
by the Paperback Division of
W. H. Allen & Co. Ltd
A Howard and Wyndham Company
44 Hill Street, London W1X 8LB

First published in Great Britain by
W. H. Allen & Co. Ltd, 1980
A Step Farther Out parts 1 and 2 were originally published as one volume.

Reproduced, printed and bound in Great Britain by
Hazell Watson & Viney Ltd, Aylesbury, Bucks

ISBN 0 352 30906 7

Dedication

For Jim Baen, an extraordinarily
good friend and editor.

Acknowledgments:

Research for this book was supported in part by grants from Pepperdine University and the Vaughn Foundation, to both of whom the author gives respectful thanks. Opinions in this work are the sole responsibility of the author.

It is obviously impossible to thank all those who have significantly contributed to a work this size. However, special thanks are due to Ejlar Jakobssen, who first encouraged me to do a regular science column; to Jim Baen, his successor as editor at *Galaxy*, who became invaluable for his topic suggestions and who unarguably improved the result; to Freeman Dyson, who generously granted ideas and interviews; to Dr. Gerald Yonas and Dr. John Penitz of Sandia Corporation; Dr. Petr Beckmann; Russell Seitz; A. E. van Vogt and Robert Bloch for their encouragement; Edmund Clay; and Larry Niven. Thanks are also due the members of the Los Angeles Science Fantasy Society, Inc., who were hardy enough to listen to my readings of these essays before they saw print.

It is also traditional for an author to thank his wife for putting up with him while he went through the emotional storms so often associated with writing. In my own case the thanks are for reasons considerably more valid than tradition.

JEP

Hollywood, 1978

Contents

Preface

The Freedom of Choice
by Larry Niven

Jerry Pournelle is out to make the whole world rich.

He's been at this for some time. Like a good many of his colleagues, Jerry was sucked into the space sciences by science fiction. He was building rockets for the government back when they had to steal the parts from other projects, and get the work done by sneaking back into the plant after clocking out. He's been building the future since I was in grade school, and he's still at it.

Of course, he would prefer to build it *his* way. Jerry has less of the ability to "suffer fools gladly" than anyone I know. (I'm not too good at that myself.)

His ambitions are impressive. In this book you'll find laid out for you several routes to a future in which the entire world is as wealthy as the United States is today . . . and that is as wealthy as any nation has been in human history. He does not intend that we should confine ourselves to Only One Earth.

Well, you'll get to that. Let me deal with another question. Do we *want* the whole world rich?

I happen to think we do, but I've heard other opinions.

Do you feel that your soul and body will benefit if you eat nothing but organically grown fruits and vegetables? You may well be right; but there's a reason why those scrawny carrots are so expensive. Without fertilizers and bug sprays the tomatoes, etc., might not come out of the ground. (Ours didn't!) Wealth lets you pay someone else to grow it. If you go the whole route, forming a commune, living as your ancestors did, eating only food you grow yourself without technological help . . . then wealth lets you go on eating after the crop fails.

More generally, the right to live as if you were poor is inalienable. What you stand to lose is the right to live otherwise. Through your laziness or your inattention or through listening to the wrong saviors, you may condemn all future generations to involuntary poverty.

Nobody can be forced to spend wealth. That applies to you as thoroughly as it applies to the Indian rice farmer or Brahmin mendicant. Either can simply ignore the wealth that Dr. Pournelle proposes to drop on his head.

Granted that there are problems. A wealthy world would aggravate the servant problem no end.

Remember when people could sell themselves into slavery in order to eat? There was a ready market, because machines did not yet compete with musclepower. Those halcyon days are gone. With no good reason to fear for their jobs, servants have already become arrogant enough that most people would rather let a machine do it.

Well, why not? In the past few decades we've developed ultra-dependable ovens, vacuum cleaners, dishwashers, washer-dryers, soaps and detergents and other specialized chemicals for tasks each of which was once served by elbow grease (and somebody else's elbows, with any luck). The controls on my microwave oven have a better memory than my mother's cook, and my mother's cook quits more often.

In an age of inflation, the price of computer capability is going *down*. Ten years from now, your chauffeur may well be a computer; and why not? It would take up less room in the car and *far* less room in the house.

Consider backpacking. Over the decades, what was once a test of survival has become comfortable. Roads carry you into the wilderness. There you carry freeze-dried food and a lightweight mummy bag and air mattress on a contoured pack with a hip belt. Naturally the trails grow crowded. The population increases, the wilderness decreases. Already people propose to put glittering solar power collectors all over perfectly good deserts, instead of in orbit, as God intended.

If four billion people could afford to buy Kelty packs and sleeping bags, a certain minute percentage would go backpacking. And the world's wilderness areas would be jammed! What happens to the original backpacker, the man who needs the solitude of an empty trail?

No sweat. If we follow Jerry's route, we'll be moving a lot of our industries into Earth orbit and beyond. We'll be mining the Moon and the asteroids, and using free fall to keep heart patients alive and to manufacture ball bearings and single-crystal whiskers and strange new alloys. Let's continue that process. Move *all* of Earth's industries into Earth orbit. Turn the Earth into one gigantic park. There'll be room for the backpackers.

Does the world need to be rich? Suppose the worst: suppose none of the money is yours. What does the wealth of a society do for you?

The last time I spoke on this subject, someone in the audience called me a "bourgeois" for the first time in my life. Do we bourgeoisie tend to overemphasize wealth? Maybe. Someone else pointed out that, if we were all to spend most of our time in meditation, in seeking out the strengths and weaknesses of our own souls, we would use very little of the world's resources.

She was right, of course. I did have to point out that one would get the same benefits from being dead; but even that isn't the point. *Choice* is what matters. You have the right to choose your profession or lack thereof, your friends, when and whether you get married, what clothes you wear, how and whether to cut your hair and shave your face or legs;

10

and whether you spend twenty-four hours a day meditating. But that right depends absolutely on your ability to walk out! If the pressure from your parents or neighbors is too much, hop on a bus and go. Change cities, if necessary. You don't have to resist the pressure to conform. There are people living exactly as you would like to. Find them!

What does it take to maintain these freedoms? Not much. Classified ads in newspapers, a nationwide telephone network, your car and a network of highways and gas stations, several competing airlines, a public police force—actually a fairly recent invention, that one.

Fred Pohl's biography speaks of another freedom—a freedom you will hopefully never need. Fred grew up during the Depression, in a society that could not yet afford Welfare. There was no bottom to failure in those days. You could starve in the street, just like in India. Far and few were those willing to claim it was good for their own souls.

Oh, there's one more freedom worth considering, for those of the female persuasion. Laws tend to pragmatism. Your legal right to be considered the equal of a man depends on physical strength being irrelevant; and that depends on machines. Women have been slaves in most societies throughout most of human history. Sophisticated contraceptives help too; they allow you to avoid compulsory pregnancy. Peasants don't manufacture contraceptives.

If you're my age (forty) or younger, you've been living in a wealthy world for all of your life. Perhaps you haven't noticed. It's time. The sources of our wealth are running out. Dr. Pournelle will show you where to go for more.

Foreword
by A. E. van Vogt

It is hard to over-praise this collection of articles on the latest developments in science, or their author. But I'm going to try.

All of these pieces appeared originally in *Galaxy*, a major science fiction magazine. When the first one was printed in the April 1974, *Galaxy*, I was motivated to write the following letter to Jim Baen:

Dear Editor: I was pleasantly surprised when I finally opened your latest issue—the April—and saw Jerry Pournelle had done a science article for *Galaxy*. This is sf magazine science fact really moving up in the world. Of all the people I have met in my life, Jerry is one of the most colossally educated in science. He had advanced training in systems engineering, psychology, physics, mathematics, logic, and political science. He is the only science degree person I have ever known who was able to explain coherently the entire Velikovsky controversy, an hour later do the same for the Atlantean legend, and then, in a major talk, describe the new, dynamic, holographic view of the structure of the brain. Jerry has Isaac Asimov's memory in a younger body, and it comes out by instant association in a similar electrifying voice. Several sentences in his article referred to new developments in aerospace and energy science that were hitherto unknown to me—and I *try* to keep up. After reading that, the future already looks brighter to me. Scientifically oriented readers will now have to add *Galaxy* to their list of publications they need to keep up with what's going on.

That was published in the July 1974, issue. Now, here we are more than four years later. During the interval, Jerry has not only written a science article for every issue of *Galaxy*; he has also become a major visiting science lecturer at universities, and has gone up in the sf literary world to become—hear this!—one of the half dozen authors in the field who have received advances of more than a hundred thousand dollars. In 1973 he won the John W. Campbell, Jr. award as the most promising new sf writer of the year. In 1977, he and his collaborator, Larry Niven, were paid an advance of $236,500 for the paperback rights of their sf novel, LUCIFER'S HAMMER.

In these articles you will find that the author is pro-technology, pro-space program, pro-interstellar exploration. And he supports these and other pro-science projects for a strange reason. He can prove that they and they alone will accomplish what the anti-science proponents want. Without space colonies, the third world is dead. Without meteorite mining, people on welfare will presently get nothing.

Read Jerry's SURVIVAL WITH STYLE and A BLUEPRINT FOR SURVIVAL. We have (according to Jerry Pournelle) a hundred years to get out into space and save ourselves. After that, because of the depletion of the necessary resources down here on the surface of the planet—necessary, that is, for getting us off and up—the opportunity will never occur again for the human race.

A science article by Jerry Pournelle has an astonishing amount of writing energy in it. Like Isaac Asimov, Jerry puts himself out where you can see who's doing the reporting. Like Isaac he knows the facts and has the formal training to evaluate and present the data.

Another comment on that training: Jerry's mother once told a group of us that as the years went by, and there was Jerry still in college, taking this degree and that degree, and that training, and that one, and that, and that ... the family began to be worried.

13

As they, and we, may see, he came out of it all right. And not only as a brain. Jerry has a tall, lean, tough body which, in its time, served in Korea, achieved a high level of skill in the graceful, muscular art of fencing, and acquired the enduring heart and lung power that comes from hiking in the mountains.

Jerry has been an aide to a mayor of Los Angeles, a practising PhD psychologist, president (the same year winning that Campbell award) of the Science Fiction Writers of America, and other achievements.

As you have now seen, I've tried to over-praise Jerry Pournelle and what he has written in this book. But it can't be done.

Introduction

I want to show you marvels. Dreams, in technicolor, with sharp edges. I want to tell you something of the wonder and excitement of science; of the birth of the universe; of black holes, and cities of the future; of how man and computer may forge between them something greater than both; of the world of energy, from garbage to outer space; of worlds transformed, and how many may direct the evolution of stars. I want to show you a world that might be made.

I also warn you: you will be asked to make some decisions. They are not decisions you can avoid; you are making them now, every day of your lives; but you may not know just how important is your generation.

We live in an age of marvels. Despite that, we feel a sense of impending doom. When I was an undergraduate I was involved with dramatic arts, and at one time was assistant director of a theater. In due course I was given original plays to read with a view to producing them.

One was memorable only for its title: "First Document of the Last Generation."

"It seems to me, then, that by 2000 AD or possibly earlier, man's social structure will have utterly collapsed, and that in the chaos that will result as many as three billion people will die. Nor is there likely to be a chance of recovery thereafter."

Thus closes a popular article by Dr. Isaac Asimov, perhaps the best-known science writer in America, written even as Neil Armstrong set foot on the Moon. Not long after the Eagle landed, there was held in Stockholm a world conference entitled "Only One Earth"—and the consensus of the meeting was that "Spaceship Earth" was very likely doomed.

We are afraid of the future. At a time when the marvels of science are spreading throughout the world, when communications satellites bring hygiene and learning to the farthest corners of the globe, we have lost confidence in science and technology; indeed, in our abilities to control our world.

The intellectuals cry "Doom!" and many of us heed the warning.

Perhaps this is as it should be. Perhaps this loss of confidence in the ability of man to master his environment is justified. Perhaps we do not "have dominion" over the Earth, and the message of the doomsayers, of the advocates of "Zero-Growth," of those who put their faith in "small systems" and "soft energies" and "appropriate technologies" is no more than good sense.

Yet it does not seem obvious, at least not to me; and surely we ought not take such a momentous decision without thought of the consequences.

For make no mistake: the consequences of anything like Zero-Growth, of abandoning the idea of progress, are real and profound, and may not be reversible. The decision may be irrevocable, through all ages of ages.

And that is the reason for this book. I frankly reject the counsels of doom. I do not do so blindly. I am as aware as anyone of the dangers we face in the coming decades. I have certainly studied the problems, and it would be foolish to ignore the warning signs; mindlessly to proceed as we have done in the past would be disastrous—but I think it no less disastrous to cry "Doom!" and abandon faith in both technology and progress.

16

Of course this book does more than examine the alternatives we face. Over the years I have written of the fascination and excitement of science, and I have hoped to convey some of the sense of wonder—the conviction that we live in a generation of wonders—that I have felt. Perhaps that is the more important part of the book; yet I cannot help thinking that it is a worthwhile effort to show that we are not faced with doom; that the West need not close down; that we will survive.

I hope to convince you that this is the most important generation that ever lived; that what this generation decides will affect man's future every whit as much as did the development of agriculture, the invention of the wheel, the harnessing of fire; indeed, perhaps, as much as the evolution of lungs.

That is no small thing.

PART FOUR:

SPACE
TRAVEL

COMMENTARY

"Halfway To Anywhere" was the first column I wrote for *Galaxy*. I haven't changed it, because (1) the numbers are still correct, and (2) it's interesting for its history: the column was written in November of 1973, and the tables were made up from slide rules and logs; archaic methods indeed! I'd never do that much work now, were I robbed of my calculators and computers.

The second chapter, "Those Pesky Belters and Their Torchships," was never printed. There were too many tables and too many numbers. A much watered-down version eventually appeared as the second column I'd ever done, and then-editor Ejler Jakobbsen gave me to understand that in future I'd not include *quite* so much detail. He may have been right; I haven't yet made up my mind. Certainly when Jim Baen took over the editor's job at *Galaxy* a lot more detail and many more equations crept back into the column.

Science writers have a problem: how much detail do we include, and how technical can we get? After all, our first purpose is to entertain; if we can't do that, there's no point in writing a column or article for the general public. On the other hand, there's buried in most of us a frustrated teacher: we want the readers to *understand* and even to be able to work these things out for themselves.

And many readers do want the gory details; especially in this era of cheap but excellent calculators and computers. Thus there's the temptation to include the equations, and a lot of tables and charts. At the same time, most adult readers did *not* grow up in the era of the calculator and are understandably a bit, uh, perhaps not afraid of, but uncomfortable with numbers and equations. Those who open a book and see a ton of equations are likely to put it back down again unless they're part of the small minority who like such things.

I've always tried to strike a balance: to write these columns so that you could follow the arguments without having to work the equations, but to give enough details to let those genuinely interested play the game themselves. I am told I've been reasonably successful at that.

Anyway, this section presents "gory details"; but you really don't have to do the math to see what's being said, and hopefully to enjoy the speculations about space travel.

Halfway to Anywhere

One of my rivals in the science-writing field usually begins his columns with a personal anecdote. Although I avoid slavish imitation, success is always worth copying. Anyway, the idea behind this column came from Robert Heinlein, and he ought to get credit for it.

Mr. Heinlein and I were discussing the perils of template stories: interconnected stories that together present a future history. As readers may have suspected, many future histories begin with stories that weren't necessarily intended to fit together when they were written. Robert Heinlein's box came with "The Man Who Sold the Moon." He wanted the first flight to the Moon to use a direct Earth-to-Moon craft, not one assembled in orbit; but the story had to follow "Blowups Happen" in the future history.

Unfortunately, in "Blowups Happen" a capability for orbiting large payloads had been developed. "Aha," I said. "I see your problem. If you can get a ship into orbit, you're halfway to the Moon."

"No," Bob said. "If you can get your ship into orbit, you're halfway to *anywhere*."

He was very nearly right.

Space travel isn't a matter of distances, it's a question of velocities. Now most space systems designs begin with rough-cut estimates of present and near-term predicted technological capabilities; and one of the best measures used in design analysis is called "delta vee." This is engineer talk for a change in velocity, and comes from the general mathematical symbol for change, the Greek letter delta or Δ. Delta vee, written Δv, is the total velocity change a ship can make.

The nice part about delta v is that for rough analysis it doesn't matter how you expend your fuel. You can burn it all up at once, or make a whole series of velocity changes: the sum of delta v achieved will be the same. Moreover, the total delta v can be calculated from the Specific Impulse (a measure of efficiency) of the fuel used and the fraction of the total ship weight that's made up of fuel. No other numbers are needed, not even total ship's weight. Given the total delta v, you can determine what kind of missions the ship can perform.

The other nice feature is that delta-v requirements for any journey in the solar system can be calculated from well known parameters: mass of the sun, masses of the planets you're leaving and going to, and the distances of the planets from the sun. There are a lot of possible refinements, but rough estimates of delta-v requirements for any minimum energy journey can be run off on a slide rule in no time.

The least costly method of long-distance space travel involves transfer orbits, sometimes called Hohmann orbits after the German architect Dr. Walter Hohmann, who first calculated the energy requirements to get from place to place in the solar system. Hohmann's book, THE AT-

TAINABILITY OF THE CELESTIAL BODIES, was published in the mid-thirties and was a very important book indeed, because it showed that space travel really was possible with chemical rocket fuels.

Unfortunately, as Willy Ley noted in ROCKETS AND SPACE TRAVEL, Hohmann's book is nearly unreadable, combining Germanic scholarly thoroughness, unfamiliar subject matter, lots of mathematics, and a terribly complex style. Despite that, his work remains important and the transfer orbits he described are the only feasible methods of getting to other planets from Earth with chemical rockets.

In Hohmann orbits, the starting planet at GO and the target planet at the time of journey's end must be precisely opposite each other with the sun between. Naturally, then, the trip begins when the target planet hasn't yet got to opposition, and these journeys can start only at certain times. The ship departs on a trajectory that carries it into a highly elliptical orbit with one end of the ellipse just touching the orbit of the origin planet and the other touching the orbit of the target planet.

The delta v required for Hohmann trips to various places is shown in Figure 12. In every case it is assumed that the starting point is not on Earth, but in orbit around Earth. The numbers were calculated for me by Dan Alderson, who programs JPL's computers and is usually concerned with real spacecraft such as Pioneer and Mariner; they're quite accurate, given the model used. For those interested, we assume the planets have circular orbits and all lie in the same plane, and use conic section approximations.

The first important number is the fly-by delta-v requirement. This assumes you just want to get close to the target, and after that you don't care what happens to the ship. In the real world, fly-by probes can be useful afterwards: the Pioneer series Jupiter probes, for example, may round Jupiter in such a way that they use Jupiter's attraction to fling them on toward other planets, or out of the solar system altogether.

There was even a possibility of a Grand Tour, in which the spacecraft approached Jupiter, Saturn, and then either flew past both Uranus and Neptune, or went direct from Saturn to Pluto, each time using the delta v gained from a close approach to one planet to get to the next. Congress wouldn't fund the Grand Tour, and that opportunity is lost for our lifetimes because it takes a special configuration of planets for the Grand Tour; but as I write this they're preparing to launch Mariner 10, a probe that will use Venus as a slingshot to send it down to Mercury, and if all goes well Mariner will arrive at Mercury about the time this is published. (Mariner is scheduled to arrive at Venus on Feb. 5, 1974, and at Mercury on March 29, 1974.)

The Pioneer probes carry the famous gold plaque with a code showing the origin of the spacecraft and line drawings of human male and female, on the assumption that someday they may be picked up by beings in another star system. Since the probe will leave the solar system with a velocity of only a few kilometers per second, and must cross trillions of kilometers before there's any possibility of its being found, we don't have to worry much about the aliens using it to track us back to Earth and conquer us; by that time, if interstellar travel is possible, we'll have it.

It happens that I was present when that plaque—called "The Praque" by the TRW technicians who built Pioneer —was invented. NASA held a big press briefing at TRW, a dog and pony show for science reporters; the NASA, JPL, and TRW scientists concerned with Pioneer described the experiments aboard, and one happened to mention that Pioneer would definitely leave the solar system forever.

One of the reporters present was Eric Burgess, who with Arthur Clarke founded the British Interplanetary Society back in the 40's. Eric became very thoughtful, and later that afternoon spoke to Carl Sagan of Cornell and some of the others in charge of Pioneer, pointing out what a unique opportunity this was to send a message to anyone "out

Figure 12

DELTA V REQUIRED FOR VARIOUS TRIPS BEGINNING AT EARTH ORBIT

TARGET	AVERAGE DISTANCE FROM SUN (KILOMETERS)	FLY-BY DELTA V (KM./SEC.)	MARGINAL CAPTURE DELTA V (KM./SEC.)	CIRCULAR CAPTURE DELTA V (KM./SEC.)
Sun		21.249		200.786
Mercury	57,900,000	5.580	11.874	13.104
Venus	108,000,000	3.555	3.905	5.470
Earth	148,000,000	3.280	3.280	3.280
Mars	228,000,000	3.661	4.320	5.535
Asteroid	300,000,000	4.378	8.320	8.320
Ceres	414,000,000	4.691	9.530	9.530
Jupiter	778,000,000	6.322	6.583	10.315
Saturn	1,430,000,000	7.293	7.691	11.143
Uranus	2,870,000,000	7.981	8.469	11.277
Neptune	4,500,000,000	8.248	8.575	11.116
Pluto	5,910,000,000	8.363	8.841	10.972
Escape	infinite	8.748		

Values for Sun are very close approach and circular orbit at surface. Value for Earth is marginal delta v needed to escape Earth's gravitational effect. Asteroid capture values are large because the asteroids have essentially no mass, and thus do not aid appreciably in an attempt to catch up with them after arriving at their orbital distance.

there." It might take a long time to arrive, but at least it was going. The idea caught on, and within a week the plaque was designed and installed.

Then, of course, came the complaints about the "dirty pictures" of nude men and women, but that's another story.

Figure 12 shows in addition to fly-by delta-v requirements, the delta v you'd need to get into some kind of orbit around the planet: the bare minimum for capture, and a circular orbit from which you could land or observe closely. You can see the numbers come out at reasonable values, except when you're trying to get very close to the sun. One important number is the Sun escape velocity. If you have that much delta v capability, you can get to other stars: anywhere, for practical purposes. It is important to note, though, that Figure 12 assumes you don't start from Earth, but from *orbit around Earth.*

Since you need 7.6 km/sec delta v to get into Earth orbit in the first place, Bob Heinlein's top of the head remark was very close to correct. Earth orbit is halfway to anywhere.*

In other words, the first step's the hard one. If you can get into Earth orbit, you can get most anywhere else. Unfortunately, the disintegrating totem poles we now use to get into orbit are just too cumbersome and expensive to make space flight routine. Worse, they use up nearly all their total delta v getting into orbit—and the rocket is thrown away, hundreds of millions of bucks into the drink.

The upcoming shuttle reusable ship will help and is sorely needed, but there's a system even better than that. The concept I'm about to describe can use old rocket boosters over and over again; in fact, the rocket motor never leaves the ground. Only payload goes up.

This magic feat is performed by lasers. The basic design

---

*Quibblers will know that you'd have to stay in the plane of the ecliptic or use a lot more energy to get out of it; and that the Galaxy itself has a very high escape velocity, in the order of 100 km/sec from here...

of the system comes from A.N. Pirri and R.F. Weiss of Avco Everett research laboratories. What they propose is an enormous ground-based laser installation consuming about 3,000 megawatts. In practice, there would probably be a number of smaller lasers feeding into mirrors, and the mirrors would then concentrate the beam onto one single launching mirror about a meter in diameter. This ground station zaps the spacecraft; the ships themselves carry no rocket motors, but instead have a chamber underneath into which the laser beam is directed.

The spacecraft weigh about a metric ton (1000 kilograms or 2200 pounds) and are accelerated at 30 g's for about 30 seconds; that puts them in orbit. While the capsule is in the atmosphere the laser is pulsed at about 250 hertz (cycles per second when I was in school). Each pulse causes the air in the receiving chamber to expand and be expelled rapidly. The chamber refills and another pulse hits: a laser-powered ramjet. For the final kick outside the atmosphere the laser power is absorbed directly in the chamber and part of the spacecraft itself is ablated off and blown aft to function as reaction mass. Of the 1000 kg. start-weight, about 900 kg. goes into orbit.

Some 80 metric tons can be put into orbit each hour at a total cost of around 3000 megawatt-hours. Figuring electricity at 3¢ a kilowatt hour, that's $150 thousand, less than a dollar a kilogram for fuel costs. Obviously there are operating costs and the spacecraft aren't free, but the whole system is an order of magnitude more economical than anything we have now.

Conventional power plants cost something like $300 a kilowatt; a 3000 megawatt power plant would run close to a billion dollars in construction costs. However, when it isn't being used for space launches it could feed power into the national grid, so some of that is recovered as salable power. The laser installation might easily run $5 billion, and another $5 billion in research may be needed.

The point is that for an investment on the order of what

we put out to go to the Moon, we could buy the research and construct the equipment for a complete operating spaceflight system, and then begin to exploit the economic possibilities of cheap spaceflight.

There are a lot of benefits to an economical system for getting into orbit. Some are commercial, things like materials that can only be made in gravity-free environments and such like. Others are not precisely commercial, but highly beneficial. For example, the power/pollution problem is enormously helped. Solar cells can collect sunlight that would have fallen onto the Earth. They convert it to electricity and send it down from orbit by microwave. That's fed into the power grid, and when it's used it becomes heat that would have arrived here anyway; the planetary heat balance isn't affected.

Interestingly enough, it's now believed that orbiting solar power plants can be economically competitive with conventional plants, provided that we get the cost of a pound in orbit down to about $20. The laser-launch system could power itself.

We don't even have to build a permanent power plant to get the laser-launcher into operation. There are a lot of old rocket motors around, and they're very efficient at producing hot ionized gasses. Hot ionized gas is the power source for electricity extracted by magneto-hydro-dynamics, or MHD. MHD is outside the scope of this article, but basically a hot gas is fed down a tube wrapped with conducting coils, and electricity comes out. MHD systems are about as efficient as turbine systems for converting fuel to electricity, and they can burn hydrogen to reduce pollution.

The rocket engines wouldn't last forever, and it takes power to make the hydrogen they'd burn—but we don't have to use the system forever. It needn't last longer than it takes to get the big station built in space and start up a solar-screen power plant.

None of this is fantasy. The numbers work. Avco has done some experiments with small-scale laser powered

28

Figure 13

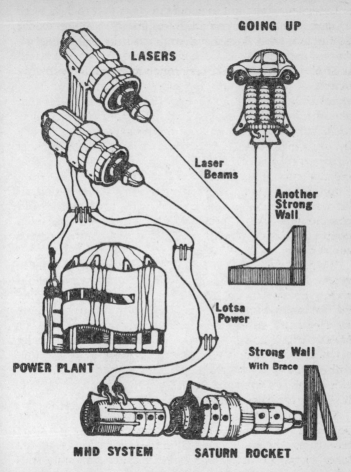

GOING UP

LASERS

Laser Beams

Another Strong Wall

Lotsa Power

POWER PLANT

Strong Wall With Brace

MHD SYSTEM   SATURN ROCKET

PERSONAL SPACE TRANSPORT SYSTEM

"rockets," and they fly. There are no requirements for fundamental breakthroughs, only a lot of development engineering, to get a full-scale working system.

Laser-launchers are in about the stage that rockets were in, say, 1953. Fifteen years and less than $20 billion would do the job: we'd have a system to get nearly anything we wanted into orbit.

That doesn't seem like very much to get halfway to anywhere.

Those Pesky Belters
and Their Torchships

The other day I got a phone call from a national magazine, and being basically kind, generous, and always in need of an excuse to stop working, I spent an hour with the reporter. He wanted a list of ten science fiction predictions that have proved out, and ten more SF writers have made but which haven't happened yet.

He wasn't at all interested in a list of science fiction predictions that just aren't going to happen. Except in rare moods neither am I, but I've just finished reading the death-knell for poly-water (it turns out poly-water isn't a new form of water at all; it's just a product of dirty laboratory glassware) and that plus the phone call got me thinking about nice ideas that just won't work no matter what we do.

One of our favorite SF themes is the "Belter Civilization," which usually seeks—and gets—independence from the colonial masters on Earth. Belters make their livings as asteroid miners, and they flit from asteroid to asteroid, slicing up planetoids for the rich veins of metal we'll presumably find in them.

In the usual story, the miners go off on long prospecting tours, leaving their families on a "settled" rock. The Belt Capital is usually located on Ceres or some other central place which may or may not have been extensively transformed; and when Belters get together, it's always in an asteroid city.

The Belters don't ever come to Earth or any other planet. Indeed, they regard planets as "holes," deep gravity wells which can trap them and use up their precious fuels. The assumption here is that it's far less costly to flit from asteroid to asteroid than it is to land on a planet or get into close orbit around one.

Another assumption, generally not stated in the stories, is that fuels are expensive and scarce, and the Belters have to conserve reaction mass; this is why, in the usual Belter story, you conserve both time and energy by never going outside the Belt. Scarce as fuel is, though, I suppose the Belters have a source of it locally or they couldn't contemplate independence. They must have fuel for their ships and energy for their artificial environments. Without those, there'd be no Belter Civilization. Even if we discover something of fabulous value in the Belt we can't operate without energy and fuel.

Those are not, by the way, the same thing. Nuclear fission reactors and large solar panels could provide enough power for a permanent Belt station, and if there were something valuable enough out there we could put a reactor onto an asteroid now. Rocket fuels are something else again. To make a rocket work, you must have reaction mass: something to get moving fast backwards and dump overboard. Unfortunately, asteroids are rock, and rocks don't make very good rocket fuel. We'll come back to what the Belters might do about that later.

For the moment, let's see how difficult travel to and in

32

the Belt is. We'll use the same measure as last time, the total change in ship velocity required to perform the mission. This is called delta v, and you should recall that a ship with a given fuel efficiency and ratio of fuel to non-fuel weight will have a unique calculatable delta v. It doesn't matter whether the pilot uses that delta v in little increments or in one big burn: the sum of velocity changes remains the same.

Similarly, various mission delta-v requirements can be calculated from the laws of orbital mechanics independent of the ship used. Figure 14 gives the delta-v requirements for getting around the Earth-Jupiter portion of the solar system. We're assuming that getting to Earth orbit is free, whether with the laser launching system I described previously, or with shuttles, or whatever, so all missions to or from Earth begin and end in orbit.

The first thing we see is that landing on an asteroid isn't much easier than going to Mars; in fact, Ceres is harder to get to than Mars. This is because not only are the asteroids a long way out, but they don't help you catch up to them; they've so little mass that you have to chase them down. Thus, once among the asteroids, you may well want to stay there and not use up all that delta v coming back to Earth.

Then, too, travel between Ceres and a theoretical asteroid 2 AU out is a lot cheaper than getting to Earth from either of them. (One AU, or astronomical unit, is the distance from Earth to the Sun and is 93,000,000 miles, or $149,500,000 = 1.5 \times 10^8$ kilometers.) It takes 8 km/sec to get to the 2 AU rock, but only 3.2 more to get from there to Ceres.

So far so good, and we're well on the way to developing a Belt Civilization. There's already a small nit to pick, though: although travel to Mars itself is costly, it's as easy to get to Mars *orbit* as it is to go from asteroid to asteroid. Thus, if a laser-launch system could be built on Mars, making travel to and from Mars orbit cheap, Mars might well become the Belt Capital.

Politics being what they are, though, perhaps the Martians (well, Mars colonists?) won't like having all those crude asteroid miners on their planet, and the Belters will have to build their own capital at some convenient place such as Ceres or the 2 AU rock, saving both their feelings and some energy. However, we've so far said nothing about how *long* it takes to get from one place to another. The delta v's in Figure 14 are for minimum energy trips, Hohmann transfer orbits, and to use a Hohmann orbit you must start and finish with origin and destination precisely opposite the Sun. You can't just boom out when you feel like it; you must wait for the precise geometry, otherwise the delta-v requirements go up to ridiculous values.

You get a launch window once each synodic period. A synodic period is the time it takes two planets, or planetoids, to go around the sun and come back to precisely the same positions relative to each other: from, say, being on opposite sides of the Sun until they're in opposition again, which is what we need for a Hohmann journey.

The synodic periods and travel times are given in Figure 15, and our Belt Civilization is in trouble again. Not only does it take 1.57 years to get from Ceres to 2 AU (or vice versa), but you can only do it once each 7 years! Travel to and from Mars isn't a lot better, either. The Belters aren't going to visit their Capital very often, and one wonders if a civilization can be built among colonies that can only visit each other every seven to nine years, spending years in travel times to do it.

By contrast, you can get from Earth to the flying rocks every year and a half, spending another year-and-a-half in transit. That's no short time either, but it beats the nine years of the Ceres-asteroid visitations.

Perhaps, though, we haven't been quite fair to the Belters. Asteroids aren't as widely separated as Ceres and our 2 AU rock. Most textbooks claim the asteroids are concentrated between 2.1 and 3.3 AU out from the Sun. We'll assume they're all in the same plane (they aren't), so the

34

Figure 14

VELOCITY CHANGE REQUIREMENTS FOR TRAVELING THROUGH THE ASTEROID BELT (KILOMETERS/SECOND)

Origin or Destination	Earth	Mars	2 AU	Ceres	Jovian Moon
Earth	7.9	9.3	8.1	9.8	6.6
Mars	5.5	3.5	7.2	8.8	6.5
2 AU	8.1	3.2	0.0	3.2	4.4
Ceres	9.5	5.0	2.9	0.4	2.8
Jovian Moon	6.6	4.4	4.4	2.3	0.0

Values above the diagonal are delta-v requirements for travel from surface to surface, except when the starting point is Earth. All values for Earth assume beginning and end in Earth orbit. *Diagonal* values are circular orbit velocities. Values below the diagonal are those required for a trip that begins and ends in circular orbit. (Example: Ceres orbit to Mars orbit requires delta v of 5km/sec, while dirtside Ceres to dirtside Mars takes 8.8 km/sec.)

Figure 15

SYNODIC PERIODS AND TRANSIT TIMES FOR HOHMANN TRAVEL

From-To	Earth	Mars	2 AU	Ceres	Moon of Jupiter
Earth	----	2.08	1.50	1.28	1.09
Mars	0.71	----	5.77	4.69	2.24
2 AU	0.92	1.17	----	7.15	3.68
Ceres	1.84	1.57	1.83	----	7.56
Moon of Jupiter	2.7	3.08	3.4	3.97	----

Transit times *below* the diagonal; synodic periods above the diagonal. All figures given in Earth years.

Belt area works out to 4.6 x 10^{27} square centimeters. The books say there are about 100,000 asteroids visible with the Palomar Eye, but we want to be fair (and make things simple) so we'll assume there are 460,000 asteroids interesting enough to want to visit, or one every 10^{22} cm^2 within the Belt. That means the asteroids lie on an average of 10^{11} cm apart, which happens to be 10^6 km or one million kilometers, about three times the distance from Earth to the Moon.

Out that far from the Sun's influence, and only going a million kilometers anyway, we don't have to use Hohmann orbits. One of the Belters' usual (science fiction) devices is the torchship, which accelerates halfway to where you're going, turns over, and decelerates back again. Orbital flight doesn't really work that way, but this will do as an approximation. Figures 16 and 17 show what it takes to do that.

Examining Figure 16 we see the Belters have a problem. Any group of two or three aren't so far apart that they couldn't go see their neighbors, using the 1 cm/sec acceleration and taking a day off to do it; but at 6 km/sec delta v you won't do it often. You could try a lower acceleration, and get to more places, but before you go far, time and energy stop you again. A cluster of rocks might proclaim independence, but that cluster will be a long way from any other cluster. We could conceivably have a number of Belter Civilizations, but hardly a single Belt-wide government, even with 20km/sec delta-v torchships.

Those torchships" present a problem anyway. Continuous boost may be easy on the passengers, but it uses up delta v like mad. The one gravity boost trip for a lousy million kilometers takes more delta v than 10 round trips to Ceres! It gets worse when you try to boost over long distances, too. Figure 18 shows some representative requirements for torchship travel between planets. Sure, the travel times are now very short, and the Belt Civilization is in business: you can get from one side of the belt to the other,

Figure 16

ACCELERATIONS, TRAVEL TIMES, AND DELTA-V REQUIREMENTS FOR CONTINUOUS BOOST TRIP OF ONE MILLION KILOMETERS

Acceleration cm/sec²	Acceleration "g's"	Travel Time days	Delta v km/sec
10^3	1	0.23	200
10^2	0.1	0.74	63.2
10	0.01	2.3	20
1	0.001	7.4	6.32
0.1	0.0001	23	2

Note: figures for very low accelerations will necessarily be inaccurate.

Figure 17

TRAVEL AT .1 CM/SEC ACCELERATION CONTINUOUS BOOST

Distance km.	Travel Time days	Delta v required
1 million	23	2
10 million	74	6.32
100 million	230	20

Note: the model is especially inaccurate for very long trips.

7 AU, in a week, so trips to the Capital, no matter where it is, are no problem.

However, trips *anywhere* are no problem if you have torchships. Why bother avoiding the 8 or 9 km/sec penalty for landing on Mars or going to Earth? The "holes" aren't very deep when you have a ship like that; certainly you can afford to go to Earth orbit, and Earth is still likely to be closer to you than the Belt Capital.

Furthermore, it's hard to see how the torchships will work. The usual explanation in science fiction is "atomic power," but that's not really the whole story. We already have, or very nearly had, an "atomic powered" rocket, NERVA, but it wouldn't do anything like *that*. NERVA is an atomic pile through which hydrogen is poured as a coolant. The pile heats the hydrogen which therefore goes aft, fast, and the rocket moves.

We could have built a NERVA engine weighing about 35,000 pounds and delivering some 250,000 pounds of thrust by the end of this decade; the development was moving very well when Congress decided to cancel the program. Incidentally, more money is annually spent on lipsticks in New York state than NERVA was costing, and any medium-sized state has more annual sales of liquor than NERVA cost over its lifetime; but it's nice that they don't waste the taxpayers' money with frivolities like space. Editorials aside, it's reasonable to assume NERVA-type ships will first explore the asteroids.

The design spec for NERVA was an I_{sp} of 800 seconds. I_{sp} is pounds thrust per second per pound of fuel used, and is thus the measure of fuel efficiency. The 800 second ship might have flown by 1980, and a second NERVA with I_{sp} of 1300 seconds was possible by 1990. Given the I_{sp} we can calculate temperatures of the exhaust gasses, and thus see what we have to handle to get really efficient rockets.

Figure 19 shows some of these—and shows we're in trouble again. NERVA used a system in which the fuel passed through pipes in the motor wall to cool that before it went into the reactor chamber, and with similar design features could handle gasses with temperatures in the 975-1300 Isp region, but the 2950 is very much an upper limit; no pile system will withstand temperatures of 50,000° Kelvin.

It follows that the best NERVA we could build won't be a torchship. With 50% of the total weight in fuel, and nearly 3000 seconds Isp, we have a ship capable of cruising between the asteroids and Earth; in fact, it can go almost anywhere in the solar system, but only in Hohmann transfer orbits. It can't accelerate continuously.

Actually, even if we could get 1,000,000° K temperatures we only have 208 km/second delta v with a ship that's 80% fuel. This is very respectable (although I don't know how we control that temperature) but it's only good for one, count 'em, one continuous burn for a million kilometers at one g. A million kilometers is not far in the solar system. True, this theoretical ship might make one of the lower acceleration boosts, but it's still no torchship as usually given to the Belters by SF writers.

Well, if NERVA can't do it, what about a "true" nuclear rocket: one that uses controlled fusion? Thermonuclear reactions surely will give enough energy for torching, won't they?

Not by themselves. There's plenty of energy in fusion, but how do you *contain* it? On Earth, with huge tokomak rings generating enormous magnetic fields, we might be able to build a magnetic pinch-bottle to hold a controlled fusion reaction; but aboard a *space ship*? If you can build something light enough to go aboard a ship smaller than the *Queen Mary* and able to contain controlled fusion, you've got a device that will change far more than the asteroid Belters: it's obviously a defense against hydrogen bombs, to begin with. We'll discuss various properties of fusion powered ships some other time; for now, it's enough to point out that they aren't the panacea we wish they were.

Figure 18

TORCHSHIP TRAVEL TIMES AND DELTA-V REQUIREMENTS FOR ONE GRAVITY ACCELERATION

Distance traveled, AU	Distance, km	Time, days	Delta v km/sec
1	1.5×10^8	2.8	2460
2	3.0×10^8	4.1	3460
5	7.5×10^8	6.5	5550
7	1.0×10^9	7.5	6530
40	6.0×10^9	18.0	15,500

Figure 19

ENGINE TEMPERATURES AND I_{sp}

Particle energy in degrees Kelvin	Exhaust Velocity of propellent cm/sec	Specific Impulse, seconds	Delta v possible, mass ratio = 2 (50% total weight is fuel)	Delta v possible, mass ratio = 5 (70% total weight is fuel)
1,000° K	4.07×10^5	415	2.8	6.6
5,000	9.55×10^5	975	6.6	15.4
10,000	1.29×10^6	1310	8.9	20.8
50,000	2.88×10^6	2950	19.9	46.4
100,000	4.07×10^6	4150	28.2	65.7
1,000,000	1.29×10^7	13,100	89.0	208
5,000,000	2.88×10^7	29,500	199	464

(K.E. = T × Boltzmann Constant, which I won't explain = ½ m v²; table assumes monotomic hydrogen fuel, with mass of 1.6733×10^{-24} grams/particle.)

Well, what *can* the Belters do? They can get around with NERVA engines, but what do they use for fuel? There isn't a lot of hydrogen out among the asteroids.

Oddly enough, there is a propulsion system that gets I_{sp} (theoretical) in the region of 20,000 seconds, doesn't have excessive real temperatures, and can be built right now. Even better, it's likely we can find fuels for it in the Belt.

This is the ion drive ship. Ion drives employ metal vapors as fuel, and the metal is accelerated by magnetic fields, not heat. If the asteroids turn out to be rich in metals, some kind of ion rocket may be just what the Belters need. It lets them get fuel from the rocks. But even with a 30,000 second I_{sp} ion drive we won't have a torchship for driving around at one g. Torching uses too much energy, and the ion drive won't develop the needed thrust anyway. A mercury vapor ion engine, for example, although very efficient, only gives thrusts of about 10^{-4} g. You get there efficiently, but it takes a long time.

No, the conclusion is obvious: with anything forseeable in the way of rockets, the Belters aren't going to develop their civilization. They won't have ships good enough to let them reach each other—and if they do get such ships, it's as easy to reach Earth as the other asteroids. With real torchships, both the Belters and the Earth Navy will have no trouble getting anywhere. I'm afraid the Belter Independence Movement is a long way off.

What with science robbing SF writers of Mars, and Venus, and now the Belters, it's all rather sad, and I've been looking for something more cheerful. I may have found it.

The Jovian Moons offer a distinct possibility for a multi-world civilization. They're respectable in size, and may well have water ices, or methane, or some other source of hydrogen on them: fuel for a NERVA engine without sending out to Mars for it.

It takes a long time to get to Jupiter's moons, so there's an incentive to stay there once you've arrived; yet it takes less delta v to get to a Jovian moon than to land on Ceres. May-

Figure 20

THE MOONS OF JUPITER

Designation (Order of discovery)	Name	Distance from Jupiter (Km.)	Radius (Km.)	Mass gm x 10^{24}	Escape Velocity (Km./sec.)
J-V	Amalthea	181,000	70	.005	.1
J-I	Io	422,000	1670	73	2.4
J-II	Europa	671,000	1460	47.5	2.1
J-III	Ganymede	1,070,000	2550	154	2.8
J-IV	Callisto	1,883,000	2360	95	2.3
J-VI	Hestia	11,470,000	50	.0018	.70
J-VII	Hera	11,740,000	10	.000015	.014
J-X	Demeter	11,850,000	7	.000005	.010
J-XII	Adrastia	21,200,000	6	.000003	.008
J-XI	Pan	22,560,000	8	.0000075	.011
J-VIII	Poseidon	23,500,000	10	.000015	.014
J-IX	Hades	23,700,000	8	.0000075	.011

Data from C. W. Allen, *Astrophysical Quantities*, 2nd ed. 1963, University of London, Athion Press. Assumes density of 3.5 grams/cm³ when no value is known.

This is the usual mass for asteroids. J-XII, XI, VIII, and IX are retrograde and highly inclined, and no calculations of velocity changes required for trips to or from them were made; their characteristics are included here only for completeness and because the data were already calculated.

be we should go there first?

The basic characteristics of the Jovian system are given in Figure 20. We've had to assume a lot of things to get these numbers, and by the time this appears some of the figures may be out of date: Pioneer will encounter Jupiter during the interval between now and when this is published. I'll be the first to cheer if Pioneer makes hash of my assumptions.

Presuming it won't, let's look at travel amongst the Jovian System. First, discard the outer satellites: they're retrograde, and take so much delta v to catch that if you can land on them, you can go nearly anywhere. The others are more reasonable, and the delta v's for travel among them are given in Figure 21. Except for Amalthea down there close to Jupiter, we can do it all with NERVA-style ships. Moreover, the travel times are very short, and you get favorable geometry for Hohmann transfer every couple of months or less. The four big Galilean Moons take about as much delta v to travel among as it took to get out there in the first place, but if they're hydrogen-rich, fueling your NERVA will be no problem.

The outer three rocks are very easy to travel among, and you could do it with a backyard rocket burning kerosene. Since those rocks are probably captured asteroids, they're as likely to be interesting as any of the others the Belters are concerned about.

So we can end on a cheerful note, saying Goodbye to the Belters, but also making ready to greet the Minister Plenipotentiary and Ambassador Extraordinary from the Jovian Moons.

Figure 21

DELTA V REQUIRED FOR TRAVEL AMONG JOVIAN SATELLITE SYSTEM

Name	Distance from Jupiter, million km	J-V Amalthea	J-I Io	J-II Europa	J-III Ganymede	J-IV Callisto	J-VI Hestia	J-VII Hera	J-X Demeter	Escape (from Jupiter)
Amalthea	.18	.007	9.4	11.9	13.8	14.5	13.4	13.3	13.3	10.9
Io	.42	7.7	1.69	5.7	8.1	9.4	9.6	9.6	9.6	7.6
Europa	.67	10.4	2.5	1.48	5.7	6.8	7.8	7.8	7.8	6.1
Ganymede	1.0	11.9	4.4	2.2	1.98	5.7	6.7	6.7	6.7	5.3
Callisto	1.8	12.9	6.0	3.7	2.1	1.63	5.0	5.0	5.0	4.1
Hestia	11.4	13.3	7.8	6.3	4.7	3.3	0.049	0.96	1.03	1.3
Hera	11.7	13.3	7.9	6.3	4.7	3.4	0.05	0.009	0.028	1.4
Demeter	11.8	13.3	7.9	6.3	4.7	3.4	0.01	0.003	0.001	1.3

All values are in km/sec. Values below the *diagonal* are delta v's needed to go from orbit around one satellite to orbit around the other, landing on neither. Values above the diagonal are surface-to-surface velocity change requirements. *Diagonal* values are the circular orbital velocities of the moons.

Escape is velocity change required to leave the Jupiter system entirely.

Ships for Manned Spaceflight

I was recently caught in an argument between Rusty Schweikert, astronaut-scientist, and Gene Thorley, Chief of the NASA Earth Resources Survey. After a while I had a sense of *déjà vu:* the man-in-space people have been fighting with the black box boys at least since 1954 when I first got involved in the space business.

The arguments haven't changed, although you'd have thought Skylab might Have ended it; but no, the black box boys still say that anything a man can do in space, instrument probes can do better and cheaper—and two instrument missions cost less than one manned mission. Meanwhile, the astronauts say that you can't learn anything from a probe that you hadn't thought to ask it to tell you, and a trained astronaut-scientist will think of new experiments while he's in space and can do them.

I won't pretend neutrality, having started in the Human Factors laboratories and put several years into work on keeping men alive in the space environment; obviously I'm prejudiced in favor of the astronauts.

Unfortunately, the black box boys have a strong point. It's going to cost a *lot* to send manned missions to the planets, and with present spacecraft, manned interplanetary travel can *never* become routine. It's not just a question of refining what we've got, either; there's a definite theoretical limit to what we can do with chemical rockets, and manned interplanetary cruises—other than spectacular one-shots to say we've done it—are beyond those limits.

Let's see why and then look at ships that *can* take man to the planets.

First we'll need a couple of basic facts about propulsion. Jim Baen and I are agreed that this chapter won't become a substitute for a textbook, and I'll keep the lecture short.

Rockets are still the only means of interplanetary travel we have, and they work by throwing mass overboard. The more mass tossed, and the faster it goes, the more thrust:

Thrust (force) = (rate of mass loss) x (exhaust velocity)

$$\text{or. } T = \frac{(M_0 - M_1) \times (V_e)}{1} \qquad \text{(Equation 1)}$$

What this says is you take the weight of the ship when you get through burning fuel, subtract that from the weight you started with, divide by time of burn and multiply by the velocity the burned fuel went aftwards from the ship; and you get thrust. Now in the foot/pound measurement system, thrust is expressed in *pounds* and sounds like a weight. It's not, it's a force, but it does have one convenient property.

One pound of thrust will just support one pound of weight in the gravity of the Earth's surface. It will accelerate that same mass at one gravity if you started in orbit. In metric systems, thrust is expressed in newtons or dynes, and if you don't understand the difference between *mass* and *weight*, they aren't easy to interpret.

However, equation one makes it easy to see that if you don't want to throw much mass overboard, you'd better get it going aft at a fast clip—that is, you need high V_e.

Now V_e happens to be related to the temperature of the burn in the rocket engine, so naturally there's a limit to how high that can get; if it's hot enough, the engine melts, and your rocket won't work so good. It should be clear, though, that the theoretical V_e obtainable with any given combination of fuel and oxidizer is a good measure of how efficient that would be as a rocket fuel.

However, for reasons I won't go into, instead of V_e most engineers use another measure of fuel efficiency known as Specific Impulse, abbreviated I_{sp}; and this is given in units called "seconds." That's not really a unit of time. I_{sp} is the "pounds of thrust obtained per pound of fuel expended per second," obviously a measure of efficiency; and

$$I_{sp} = \frac{V_e}{g} \qquad \text{(Equation 2)}$$

where g is the acceleration of gravity: 980 cm/sec/sec in the metric system and 32 ft./sec/sec in the English.

Now we have a measure of fuel efficiency, but we don't know what we need for interplanetary travel. As I've shown before in these columns, one of the most convenient figures to look at is "delta vee," meaning the total change in velocity we can get from a ship if we burn all the fuel in it. Delta v is convenient because it doesn't matter if you burn all the fuel at once, or keep turning the motor on and off; and furthermore, we can calculate the delta v needed for various space missions without knowing anything about the ships at all. Big or little, it takes the same delta v to get into orbit, or to go from here to Mars.

We're almost done with the rocketry basics, but we need one more equation:

$$\text{Delta } v = V_e \, \log_e\left(\frac{M_0}{M_1}\right) \qquad \text{(Equation 3)}$$

What this says is that the total velocity change you can get from a rocketship can be found by knowing the exhaust velocity of the fuel burned, and the ratio of the mass when you started to the mass when you finished the mission. Log-base-e is the "Natural log" of that ratio, and if you don't understand logs, don't worry about it. It's tabled in handbooks or given by scientific pocket computers.*

Early chemical rockets used fuel/oxidizer combinations with I_{sp} of under 200 seconds. That means that at a mass flow rate of 1 pound a second, the rocket could lift about 200 pounds against gravity. Recall, though, that this is the total weight lifted, including the fuel to be burned in the next seconds, etc., and you'll see it's not so good after all.

Incidentally, if you'll look at the three equations you'll see why liquid-fuel rockets start lifting slowly and get faster and faster as they rise. The mass-flow stays the same, but they're burning fuel and getting lighter all the time. Since the thrust hasn't changed, but the mass it has to lift is decreasing, the acceleration the motor can impart to the rocket is increasing all the time.

Rocket chemists worked very hard to get higher I_{sp} (and thus higher exhaust velocities), and now the best solid-fuel rockets have I_{sp} in the order of 250 seconds. Meanwhile, liquid-fuel rocket motors were developed to give even higher specific impulses, and by using liquid oxygen (LOX) and routing that around the motor to cool it before it was burned, higher burning temperatures were achieved. Eventually, LOX made I_{sp} of 300 nearly routine. More exotic fuels were employed, including liquid hydrogen with LOX, and finally hydrogen and fluorine.

---

* An easier way to calculate delta v is to send $3.00 to RAND CORPORATION, 1700 Main St., Santa Monica, Calif. 90406 and ask for a "Rocket Performance Computer." It's a circular slide rule developed by Ed Sharkey, and it comes with a book of instructions. You enter with I_{sp} and mass ratio and it gives delta v directly.

TABLE ONE

Best Delta v for chemical rockets.

$$I_{sp} = 400 \quad V_e = 3.9 \text{ km/sec}$$

Mass Ratio	% Fuel	Delta v (Km/sec)
2	50%	2.7
4	75	5.1
5	80	6.3
10	90	9
20	95	11.6

However, with the best chemical fuels, the theoretical maximum I_{sp} is no more than 400. Actually, no one has got *that* yet; and it's certain that no chemical rocket will do better.

I_{sp} of 400 is an exhaust velocity, at best, of 3.9 kilometers/second. Let's plug that into equation three for various mass ratios and see what happens. (Table One.)

It takes about 8 km/second to get into Earth orbit, which means the best chemical rockets have to be almost 90% fuel simply to get to orbit in a single stage; but let's assume we start in Earth orbit.

To get to Mars requires about 5.5 km/sec velocity if you *start* in Earth orbit—and it takes that much again to get back. Clearly, if we're to carry any payloads to Mars, we need refueling out there.

Equally clearly, we've got a problem, because how do we ferry enough equipment to set up a fueling station on Mars? About 80% of our payload put into Earth orbit gets burned before we reach Mars.

To make a round trip, which means carrying the fuel to come home with, we have to put 20 pounds into Earth orbit for each pound making the trip from Earth to Mars and back again. Even one way takes a factor of 4 to 1, not impossible, but expensive enough. Still, we could imagine commerce in which 4 tons of fuel were burned for each ton of payload delivered.

Unfortunately, minimum energy transfers (and those are the only kind possible with chemical rockets; anything else required delta v in the tens to hundreds of km/sec) take *time*: 260 days to Mars.

People eat, drink, and breathe. With the best recycling we're going to have consumables aboard the ships—and the recycling systems are massive, too. Let's be generous and say we hold things to 5 pounds a day per passenger, and start with people who weigh 200 pounds each. That's 1500 pounds of passenger and consumables—and 6000 pounds of fuel for his one-way trip.

If we can't refuel at Mars and have to carry our return LOX and hydrogen out with us, it's far worse. You need 150,000 pounds of fuel in Earth orbit for each round trip passenger. Clearly, that's not a commercial proposition. Chemical rockets can give us commerce with Mars only if we can build a Mars orbiter and fueling station, and even then it's marginal: and remember, we've given chemical rockets the best possible performance. It's not good enough.

What are the alternatives, then? We have several, one of which could have been developed by the end of this decade.

How would you like to have had a 10 person scientific mission leave Earth orbit in June, 1979, to arrive at Mars 227 days later; orbit Mars for 48 days, then head for Venus; on the way to Venus, encounter the asteroid Eros and stay near it for a day or so; then go on to orbit Venus for 55 days, and finally, 710 days after it departed, return to Earth orbit?

It could have been built. I have a model of the spacecraft that could have carried that mission. It employs a stabilized main section, and a counterbalanced rotating crew-quarters section to give about 10% gravity; it carries plenty of scientific instruments, and a small nuclear electric power plant; and by making rendezvous in Mars orbit with an expendable fuel pod, PILGRIM could have sent down a manned Mars lander.

The model was built by MODEL PRODUCTS of Mount Clemens, Michigan, and sold for about 5 dollars under the name PILGRIM OBSERVER. The engineering and celestial mechanics of the model and its mission were very well worked out—and we really could have built it for a 1979 flight. The engine employed was an atomic rocket called NERVA.

Unfortunately, neither the model nor the engine are available any longer, and for the same reason: no public interest. MPC took PILGRIM out of production at about the same time as Congress cancelled the budget of Project NERVA. This was just after Apollo 15, when people lost interest in space, and I was involved in trying to save NERVA: involved to the extent of writing some columns in daily papers, and furnishing the House Science Committee with data. But despite my efforts, which weren't important, and those of Congressman Barry Goldwater, Jr., which were very important, NERVA died.

Ironically it died a great success. It had been ground tested and found to work fine.

NERVA works like the "atomic rockets" of the better science fiction writers of the 40's and 50's. Basically, it's a nuclear reactor with a rocket nozzle at the end: you squirt fuel, say hydrogen, through the reactor; it gets hot; and out it comes, fast, to propel the ship.

Now NERVA didn't burn any hotter than the best chemical rockets; in fact, some chemical rockets operate at nearly 3000°, which is better than NERVA's design specs called for. However, V_e, which is what you want to max-

52

imize, depends not only on the temperature of the reaction, but also on the molecular weight of what you're throwing overboard.

Hydrogen burning in oxygen produces water, with molecular weight of 18. Even hydrogen and fluorine give off HF with a weight of 10. (It's also rather corrosive, since the least moisture converts it to hydrofluoric acid.) But NERVA squirted out molecular hydrogen, and that has a weight of only two.

The best tested I_{sp} for NERVA was 650. The designers of PILGRIM assumed they'd get 850 by 1978, and that was reasonable. Most engineers now think NERVA-type craft can get I_{sp} of 1200. Let's plug *those* into equation three and see what we come up with.

TABLE TWO

Delta v from NERVA Ships
(Km/second)

I_{sp}	Mass Ratio		
	2	4	5
850	5.8	11.5	13.4
1200	8.2	17.1	19.7

Even with the lower figure, you can get a round-trip to Mars at a mass ratio of 4. With that kind of capability, commerce between the planets becomes economically feasible. It's still expensive, but given some kind of system to get materials into orbit at either end, it's more than just possible—it becomes likely.

There's another alternative to NERVA, though, and it too was studied extensively before it was abandoned. It was called ORION, and on first description it seems like the most unlikely method of space travel you ever could devise.

ORION was also known affectionately as Bang-Bang. It worked very simply: you take a *big* ship, and on the bottom you put a very thick metal plate. You hang the rest of the ship in such a way that there are a lot of shock absorbers between the base plate and the ship itself.

Then you set off an atomic bomb underneath the ship.

Believe me, the ship will *move*. When it starts to slow down, you fling another atom bomb down below and detonate it. You keep doing that until you've got enough velocity to get where you want to go.

Silly as it sounds, ORION would have worked. There were some problems. Obviously that base plate and suspension system had to be carefully designed. You probably wanted a small shielded compartment for the crew and those things that couldn't take hard x-rays, and a larger compartment, unshielded, for the rest of the cargo. None of this is all that difficult.

Another problem with ORION was coupling the energy from the bomb to the ship. Atom bombs put out a lot of x-rays and neutrons and heat, and of course once out of the atmosphere there's no blast at all. But even that problem was solved: you have to put something between the bomb and the ship, something that will absorb energy from the bomb and whap! the bottom of the ship to keep it moving: something like styrofoam, for example, which looks as if it would work despite its unlikeliness.

ORION works better from orbit, but it could lift from Earth—if it weren't for the Treaty of Moscow that prohibits surface detonations of nuclear weapons, and if you weren't worried about the possible fallout.

It has been calculated that ORION would put 5 million pounds in orbit, or land 2 million plus pounds on the Moon—and do it in one whack. That's enough for a fair-sized colony's consumables and machine tools.

Of course we won't use ORION to launch from the Earth's surface, but there's nothing wrong with using it from Earth orbit to plant Lunar and Martian colonies; it's the most efficient and cheapest form of space transportation known, believe it or not.

Moreover, ORION works us toward something even cheaper. The problem with ORION at the moment is that you're blowing off a kilo or so of weapons-grade U-235 with each bang, and that stuff's not cheap. *Aviation Week* and a few other publications have been hinting that fusion bombs with laser trigger are either already or about to be developed: with these, you don't need a U-235 primary, you just have a hot laser zap some tritium or deuterium. It's more bang for the buck, and it would power ORION nicely.

Dr. Greg Benford has also described a system that would be even cheaper: you have a big power source on the ship, say a small fission reactor. That feeds a *big* laser. Pellets of tritium or deuterium are ejected below the ship, and zapped from the laser, producing fusion to drive the ship. It's the ORION principle again, carried to its most efficient extreme.

When you've got ships like that, you don't even talk about I_{sp} and exhaust velocities, and the mass ratios are actually fractional—that is, the ship that arrives weighs more than the fuel expended to get there.

All these ships were once seriously studied. Now, it's only in universities and among science fiction people that they're mentioned, and even there most don't take ORION-type ships very seriously. Yet any of these ships could have given us the planets—and until either the NERVA or the ORION principles are exploited, the black box boys have won.

Man can dominate near-Earth space using the shuttle and laser-launchers; but until we go beyond chemical rockets, interplanetary space will belong to unmanned probes.

Life Among the Asteroids

One of science fiction's biggest problems is consistency. Whenever we make an assumption, it's not enough simply to leave it at that; to be fair to the reader, the SF writer should also see what that assumption does to everything else.

This was brought home to me when Jim Baen called to ask for a column on "What happens if we get an *economical* space drive?" The result was not only the column, but the cover story for the issue. ("Tinker," included in HIGH JUSTICE, Pocket Books, 1977).

The problem is more complex than it sounds. In fact, until we have some idea of what *kind* of space drive, there's no real answer at all.

For example: let's suppose we have a magical space drive in which we merely turn on an electric motor and "convert rotary acceleration to linear acceleration." The Dean Drive, remember, was supposed to do just that.

Incidentally, the Dean Drive wasn't suppressed by big corporations, as I've heard some fans speculate. I am personally acquainted with two men who were given large sums by aerospace companies and instructed to buy the drive if they saw any positive results whatever in a demonstration.

After all, if the thing worked just a little bit, it would be worth billions. Think what Boeing could do with an anti-gravity machine! But, alas, no demonstration was ever given, although the prospective purchasers had letters of credit just waiting to be signed.

However, couldn't we simply assume that it will work and write an article about the resulting space civilization?

No. The discovery of a "Dean Drive" would mean that every fundamental notion we have about physics is dead wrong. It would mean a revolution at least as far reaching as Einstein's modification of Newton. An anti-gravity device like that would have consequences reaching far beyond space drives, just as $E = mc^2$ affected our lives in ways not very obviously associated with the velocity of light.

This doesn't mean that "Dean Drive" systems are impossible, of course. It does mean that looking at their implications is a bigger job than I want to take on in a 5000 word chapter.

Jim's question was, "What happens if we have something that gives one gravity acceleration over interplanetary distances at reasonable costs per ton delivered?" Part of the question is easy to answer. Now that I've got my Texas Instruments SR-50 (by the way, they've come out with a really marvelous device called the SR-51, and I hate them) I can run off a couple of tables to show what we could do with such a system.

The figures in Figure 22 assume you accelerate halfway, turn end for end, and decelerate the other half, so that you arrive with essentially no velocity. The numbers aren't ex-

Figure 22

TRAVEL TIMES AND DISTANCES AT ONE GRAVITY ACCELERATION

Distance	Time (Hours)	Velocity Change (Delta v) Kilometers/second
380,000 km (Earth-Moon)	3.5	122
1 AU* (Earth-Mars, Venus)	68	2,421
3 AU (Earth-Asteroid)	119	4,194
9 AU (Earth-Jupiter, Saturn)	206	7,265
50 AU (Earth-Pluto)	485	17,123

* AU = Astronomical Unit (Average distance from Earth to Sun)
   = 150 million kilometers.

Figure 23

HOW FAR CAN WE GO AT ONE GRAVITY?

Time of Boost	Velocity Reached	Distance Covered
1 hour	35 km/sec	63,500 km
1 day	850 km/sec	1/4 AU
1 week	5,930 km/sec	12 AU
1 month	25,000 km/sec	220 AU
1 year	300,000 km/sec (Speed of light)	1/2 lightyear

act, because I haven't accounted for the velocities of the planets in their orbits—but after all, Pluto is moving about 5 kilometers a second, and Mercury about 50, and when you're playing with velocity changes like these, who cares about the measly 45 km/sec difference between the two? For shorter trips the effect is even less important, of course.

The numbers are a bit startling if you're not familiar with them. Twenty days to Pluto? They won't surprise old time Sf readers, though. A full gravity is a pretty hefty acceleration. If you don't bother with turnovers but just blast away, the results are given in Figure 23, and they're even more indicative of what one gee can do.

Of course, long before you've reached light-speed at the end of a year, you'll have run into relativistic effects. Your ship gets heavier and your acceleration drops off. I don't care how good your drive is, maintaining a full gee for a year is going to take *work*. We're only concerned with the solar system, though, so we can ignore trips longer than a month and avoid relativity altogether.

Now we can write the article, right? Wrong. The problem is that last column in Figure 22. Just how do we expect to get delta v as big as all that?

Let's illustrate. As we've shown already, delta v can be calculated from mass ratio and exhaust velocity. (If you came in late, take my word for it; we'll get past the numbers pretty quickly.)

Now you could hardly call a drive *economical* if the mass ratio were much worse than, say, three, which means that if you start with 1,000 tons you'll arrive with 333. What, then, must our exhaust velocity be to make the simple trip from Earth to Mars?

It's horrible. About 2,204 kilometers/second, and what's horrible about that is it corresponds to a temperature of 50 million degrees Kelvin. The interiors of stars are that hot, but nothing else is.

Just how are we going to *contain* a temperature like that?

* * *

One answer might be that we'd better learn how; fusion power systems may require it. OK, and the fusion boys are working on the problem. However they solve it, we can be sure it won't be anything small that does the trick.

It's going to take enormous magnetic fields, superconductors, heavy structures, and a great deal more. After all, nothing material can hold a temperature like that without instantly vaporizing, and even containing the magnetic field that holds that kind of energy is no simple job.

Let's assume we can contain fusion reactions, though. We know immediately that energy is going to be no problem for our interplanetary civilization. With plentiful energy we'll find that a number of our other problems vanish.

There won't be many "rare" materials, for example; if they're rare and valuable enough, we'll simply *make* them out of atomic building blocks. Of course it may be cheaper to go find them somewhere, such as on Mars or among the asteroids, but we'll always be up against competition from the transmuters.

Life on Earth, at least among the people of the high-energy civilizations, will change drastically. Pollution will cease to be a problem (unless the fusion plants themselves are polluters, which isn't impossible). The Affluent Society will be with us, and possibly so will be regulations and rules, bureaucracy, and all the other niceties of a universal middle class.

All this comes as a result of assuming our space drive. More central to our immediate topic is the fact that the ships will be quite large—*Queen Mary* or supertanker size, not one-man prospector jobs. Someone is going to have to put up a lot of capital to build them, and it's not likely to be the Bobbsey Twins and their kindly uncle building ships in the back yard.

Only governments or very large international corporations will be in the space ship operating business, that's

60

for sure. Thus there have to be profits in interplanetary travel. Not even governments will build more than one of these ships simply for scientific reasons. There's got to be commercial traffic.

Next, there's a technological problem: assuming we have fusion power and a method of getting electricity from it doesn't necessarily give us a space drive. Contrary to the notions of a lot of high school science teachers back in the 40's, rockets don't "need air to push against"; but the rocket exhaust certainly does need something on the rocket to push against.

What can that be? Perhaps some kind of magnetic field, but an open-end fusion system is at least two orders of magnitude harder to build than a "simple" system for generating electricity. It's one thing to take 50 million degrees and suck electricity out of it, and quite another to use that as a reaction drive.

Perhaps I'm not sufficiently imaginative, but for all these reasons I decided to shelve the one g system and design the article and story around something much simpler. In fact, if we had the electric power system, we could build these ships right now.

Ion drive systems solve the "something to push against" problem by shooting charged particles out the back end. The ship is charged, the particles are charged, and they repel each other. You can get very high exhaust velocities, in the order of 200 km/sec, with ion systems. They're among the most efficient drives known.

The trouble with present ion drives is that electricity costs weight. As an example, a currently useful system needs about 2100 kilowatts of power to produce one pound of thrust. Since the power plant weighs in the order of four tons, the total thrust is not one g, but about 1/10,000 of a gravity.

It works, but it's a little slow getting there. Not as slow as you might think: it would take about 140 days to go a full AU, and your ship would reach the respectable speed of 12 km/sec. Still, it's hardly interplanetary rapid transit.

Suppose, though, we had a fusion system to generate the electricity. It would undoubtedly weigh a lot: let's say 1000 metric tons, or about two million pounds, by the time we've put together the fusion system and its support units. We'd still come out ahead, because we'd have lots of power to play with. Assuming exhaust velocities of 200 km/sec, which we can get from present-day ion systems, we'd still have quite a ship.

She wouldn't be cheap, but it's not unreasonable to think of her as on a par with modern supertankers. She wouldn't be enormously fast: I've worked out the thrust for a ship massing about 100,000 tons with that drive, and she'd get only a hundredth of a g acceleration. Still, a trip from Earth to Ceres would take no more than 70 days, and that includes coasting a good part of the way to save mass.

A world-wide civilization was built around sailing ships and steamers making voyages of weeks to months. There's no reason to believe it couldn't happen in space.

IBS *Agamemnon* (Interplanetary Boost Ship) masses 100,000 tons as she leaves Earth orbit. She carries up to 2000 passengers with their life support requirements. Not many of these will be going first-class, though; many will be colonists, or even convicts, headed out steerage under primitive conditions.

Her destination is Pallas, which at the moment is 4 AU from Earth, and she carries 20,000 tons of cargo, mostly finished goods, tools, and other high-value items they don't make out in the Belt yet. Her cargo and passengers were sent up to Earth orbit by laser-launchers; *Agamemnon* will never set down on anything larger than an asteroid.

She boosts out at 10 cm/sec², 1/100 gravity, for about 15 days, at which time she's reached about 140 km/second. Now she'll coast for 40 days, then decelerate for another 15. When she arrives at Pallas she'll mass 28,000 tons. The rest has been burned off as fuel and reaction mass. It's a respectable payload, even so.

The reaction mass must be metallic, and it ought to have a reasonably low boiling point. Cadmium, for example, would do nicely. Present-day ion systems want cesium, but that's a rare metal—liquid, like mercury—and unlikely to be found among the asteroids, or cheap enough to use as fuel from Earth.

In a pinch I suppose she could use iron for reaction mass. There's certainly plenty of that in the Belt. But iron boils at high temperatures, and running iron vapor through them would probably make an unholy mess out of the ionizing screens. The screens would have to be made of something that won't melt at iron vapor temperatures. Better, then, to use cadmium if you can get it.

The fuel would be hydrogen, or, more likely, deuterium, which they'll call "dee." Dee is "heavy hydrogen," in that it has an extra neutron, and seems to work better for fusion. We can assume that it's available in tens-of-ton quantities in the asteroids. After all, there should be water ice out there, and we've got plenty of power to melt it and take out hydrogen, then separate out the dee.

If it turns out there's no dee in the asteroids it's not a disaster. Shipping dee will become one of the businesses for interplanetary supertankers.

Thus we have the basis for an economy. Whatever people go to the Belt for, they'll need goods from Earth to keep them alive at first. Later they'll make a lot of their own, and undoubtedly there will be specialization. One rock will produce water, another steel, and yet another will attract technicians and set up industry. One may even specialize in food production.

Travel times are long but not impossible. They change, depending on when you're going where. It costs money to boost cargo all the way, so bulk stuff like metals and ice may be put in the "pipeline": given enough delta v to put the cargo into a transfer orbit. Anywhere from a year to several years later the cargo will arrive at its destination. If there are steady supplies, the deliveries are quite regular after the first long wait.

Speculators may buy up "futures" in various goods, thus helping capitalize the delivery system.

People wouldn't travel from rock to rock much. Thus each inhabited asteroid will tend to develop its own peculiar culture and *mores*. On the other hand, they will communicate easily enough. They can receive educational television from more advanced colonies. They can exchange both technical and artistic programs, and generally appreciate each other's problems and achievements.

What kind of people will go out there? Remember that life on Earth is likely to be soft: those going out will be unhappy about something. Bureaucracy, perhaps. Fleeing their spouses. Sent by a judge who wants them off the Earth. Adventurers looking to make a fortune. Idealists who want to establish a "truly free society." Fanatics for some cult or another who want to raise their children "properly."

All this begins to sound familiar: something like the colonial period with elements as late as just before WW I.

On the other hand, the "frontier" conditions will be so different from Earth that the Belters may not be too concerned with Earth. What Earth does about them is another story.

Given fusion power, Earth could go either of two ways: fat and happy, ignoring the nuts who want to live on other planets and asteroids; or officious, trying to govern the colonies, and sending up Air Force or Navy ships to enforce edicts set down by bureaucrats who've been outside once for a month and didn't like it.

Obviously there's a story or two in either alternative.

What kind of government will evolve if the rocks are left to themselves?

Well, each might *seek* independence, but they wouldn't *be* independent. They'd depend entirely too much on commerce. Given the enormous investments required to build the ships that carry that commerce, they'd depend on big monied interests, whether private or government.

The outfits that control the shipping will make most of the rules, then. They might not reach down into the colonies themselves to spell out laws and regulations, but the big decisions will be theirs. If we envision several large competing companies getting into the act, we can envision more Belt freedom through exploiting that competition.

The corporations themselves will have to set up some kind of corporate "United Nations," simply because you can't do business without enforcement of contracts, reasonably stable currencies, and the like. Their system may or may not be influenced by pressure from Earth—depending on how much Earth even cares.

There are probably other futures that can be built up from ships of this kind, but that is one reasonably consistent picture of life among the asteroids.

I think I might like it.

Editor's note: the story "Tinker," which was based on this chapter, was nominated for a Hugo for Best Novelet of the Year.

What's It Like Out There?

I'm writing most of this in a hotel room in Toronto, which is a lovely city but no place to be if you're alone on a Sunday night, and especially no place to be if you're from California: jet lag keeps you from getting to sleep at a nor-mal hour, and the Provincial Police keep you from finding an open tavern. . .

It has been aninteresting day. I've just taken part in a Canadian TV program called "The Great Debate." The is-sue was, "Resolved: space research is a waste of time and money." Anyone who doesn't know which side I took shouldn't be reading this book. Anyone who believes I lost the debate hasn't been reading it very carefully.

I slaughtered the poor chap. It helps that my opponent, John Holt, who is a charming fellow with a distinguished record in education, chose such a silly proposition to de-fend. It is trivially easy to show that space research has pretty well paid for itself already. I chose, in fact, to assert a new proposition: that the space program is the most important activity, excluding religion, in human history.

They tape "The Great Debate" in bunches, and prior to my own I watched another: Max Lerner and Toynbee's successor at Cambridge debating the proposition that Western Civilization is in a state of irreversible and imminent collapse. As I listened it came to me that their whole conversation was irrelevant. It was as if a pair of very distinguished and learned professors in Paris were debating the same subject in 1491, unaware that this Genoese nut was making application to the Queen of Spain for a small fleet . . .

And of course I said as much in my own debate, and added that very probably in Iceland a few centuries earlier someone had won in a debate of, "Resolved, the voyages of Leif Ericson are a waste of time and money." To which Mr. Holt replied that the New World was accessible to the average family, while space never would be; that space would be restricted to scientists, astronauts, and military officers, a chosen few. The general public would never be able to go. He didn't say why.

Now at first the New World was pretty well inaccessible to anyone who couldn't get Queen Isabella to hock the crown jewels, and space is in the same situation at present; but just as the Americas were soon open to workers, farmers, administrators, soldiers, adventurers, some qualified, some merely desperate, some sent as sentence of courts, space will, probably well within my own lifetime, be open to large numbers—at least if Mr. Holt doesn't get his way. The only real question is how and when.

The how is simple technology. Shuttles will help a lot. Eventually, I trust, there will come the laser launching systems I described earlier, which can put up privately owned capsules, the equivalent of the covered wagon. There will be O'Neill Colonies—which Mr. Holt particularly hates; Luna bases; asteroid mining and refineries; Mars colonies; possibly Enceladus and the Mars-forming Project; all these and more are in the cards and there's no reason to suppose they'll be restricted to super-heroes.

The when is a little harder to predict, but in fifty years for certain. It didn't take that long to get colonies established in the New World.

So what's it going to be like to live out there?

* * *

Well, first let's take the O'Neill Colony, which is a huge cylinder in space. NASA figures we could have the first one before the year 2000 if we wanted it, and there are good numbers to show that it would pay for itself within a few years after its establishment: it can sell power to Earth, as well as serve as a base for extensive space manufacturing —and there are plenty of things that you can manufacture *only* in space.

The colony will be quite large, say a cylinder three kilometers in diameter and ten to twenty kilometers long. Windows run the length of it to let in sunlight. Under the windows is land, ordinary dirt, with hills, streams, buildings, and such like. The whole thing rotates to give artificial gravity. Let's suppose the medical people have determined that a tenth of an Earth gravity is sufficient for long-term health; that means our cylinder rotates at .026 radians a second or .25 revolutions per minute.

A colonist standing on the ground and looking up through a window above will see the stars swinging past once each four minutes. He'll also see his neighbors' fields and houses hanging in space above his head, which can be disconcerting until he gets used to it, after which it won't seem any stranger than seeing mountains in the distance.

Life in the O'Neill colony may be a bit strange, but it has its compensations. If the colonist is a farmer, he'll never have to worry about the weather. There won't be any rain —he (or someone else) will have to irrigate—but on the other hand there won't be floods, storms, or droughts (so long as the engineers keep the watermakers going). He will be able to calculate *exactly* how many hours of daylight his crops will get for the entire growing season. The only weeds and insects he'll encounter will be those brought aboard by the ecology teams.

Actually, one suspects a few pests will come along as stowaways.

Imagine the town meeting after the sparrows have got loose. One faction wants them left alone. They're cute. Another advocates shotguns. Still another abhors guns, but is willing to send to Earth for a supply of sparrowhawks. After four hours of shouting the council sets the matter aside for another day. Personally, I'd vote for sparrowhawks ... beautiful!

Machinists and mill workers will find their work little different from Earth, except that everything weighs only 10% as much. For production runs the colony probably has computer-controlled lathes and milling machines, but for one of a kind items the machinists will have to do the work. There will undoubtedly be doctors and storekeepers and librarians and tailors, and, alas, lawyers, none of whose business lives will be all that different from what they would be like on Earth.

But after working hours things get more exciting. No freeways; no cars. No subways, either. In 10% gravity the simplest means of transportation is to fly with artificial wings. There might not be any other form of transport besides walking. Why should there be? (Well, for heavy hauling you might want a few electric trucks, but surely there's no need for any individuals to own cars or trucks.)

If flying is the usual transport, grocery shopping will be like New York City, where you buy a few items a day as you need them, rather than like California where you buy bags and bags once a week and transport them in a car.

Flying also means that everyone in the colony is accessible to everyone else; every place is easily accessible to anyone wanting to get there. This can drastically change the sociology. Houses will probably have roofs, not to keep the rain out, but to keep the neighbors from looking in. The house need not be anything more than a visual screen: it doesn't have any weather to control.

What all this does to the colony's mores isn't really predictable. (Who, after all, could have prophesied drive-in movies from the first automobiles?) There's little privacy. Parents will know pretty well what their teen-age kids are doing. Whether this will make pre-marital sex more or less common isn't obvious, at least not to me. It depends partly on geography, I suppose: will there be any secluded places, dark and cozy? Dark comes when the windows are closed for the night, of course; the Sun only sets when the colony wants it to. Daylight saving time is silly in an O'Neill colony, because if you want more daylight, you simply program the window blinds to give it.

It may be that parents won't care much where their children are. There won't be any dangers in the colony; one presumes that airlocks to the outside and the like are controlled against accidental use, and also that there won't be many incompetents in the community—at least not *that* incompetent. There remains the problem of crime.

It's hard to imagine jails in a space colony, although I suppose they could be built. It's hard to imagine space muggers in the first place, or that the colonists would put up with them. They might be enslaved to the community. The cost of shipping an unwanted colonist back to Earth would be slightly colossal. On the other hand, the environment is fragile enough that you certainly don't want anyone wandering around harboring burning resentment against the colony—especially not if he has suicidal tendencies. It would be all too easy to take a number of others along in a spectacular suicide. If you couldn't send them back, what would *you* suggest?

We can presume, then, that the environment is *safe*; free of most of the dangers we live with here on Earth. Now in England the custom of dinner parties grew up only after Sir Robert Peel invented police; prior to that no one in his right mind went *anywhere* after dark, and when you visited friends you stayed at least for the night. When the Lon-

70

don Police made the streets comparatively safe it became possible to visit for the evening and go back home for the night. Such factors will affect the colony patterns of friendship too.

On the other hand, there are dangers that we don't worry about here. The most significant would be leaks. It would take a very large leak to affect the colony, of course. Small ones would be costly (air isn't cheap when it has to be taken to orbit) but easily repaired before anyone felt their effects. Still, it seems reasonable that there would be a few major airtight structures, shelters into which the colonists could crowd in the event of a major break in the pressure hull.

An interesting life, with kids learning to fly at an early age. I suppose when a parent tells a teenager he's grounded, he'll mean that quite literally.

* * *

So what do you do in such a colony? Well, what do you do now? It's simple enough to sit at home and watch TV whether you're in New York or Earth orbit. Some recreations won't be possible. No backpacking trip through the wilderness. Probably no sailboating: no wind, even if there's a lake. There may be fishing, but certainly no hunting.

On the other hand, there'll be cultural activities not available on Earth. Flying, of course; real flying, not dangling from an oversized kite, but man's ancient dream of flying like a bird. Aerial acts will probably become an art form, possibly involving a large portion of the colony population. There can also be aerial ballet, with and without wings. Up in the center of the cylinder there's no gravity. Zero-g areas are easily accessible.

Games can be strange. With that large radius and slow rotation rate, the colonists won't easily be able to tell the difference between their artificial spin gravity and the real thing: not, that is, until they begin throwing things. As soon as you throw something, say a baseball, you'll know you

71

don't have normal gravity. The ball's trajectory will be strange, and it will depend on which direction you threw it in. You'll also be able to throw the ball a very long way, so far that baseball may require much larger teams to cover the huge playing field.

In fact, any projectile motion is affected. Obviously, in one-tenth gravity you can throw a ball (or a javelin or a wrestling opponent) ten times as far as you could on Earth. A javelin-throwing athlete who can manage 285 feet on Earth would get 2,850 feet, over half a mile, in the O'Neill colony gravity. Broadjumpers would also do well.

However, there's a problem. When you loft a thrown object in centrifugal gravity, you increase the time of flight; and the ballistics become strange indeed, due to an effect called the Coriolis Force. What happens is this: from the viewpoint of an observer inside the spinning object, the "gravity" is radial. Objects dropped tend to fly directly away from the center. They fall toward the "floor," and in 10% gravity as we have here, they fall rather slowly. It takes two full seconds for something to drop two meters.

While the object is falling, the "floor" is moving, so that the dropped object does *not* strike the spot directly under it. The discrepancy is related to the rate of spin and the radius of the spinning craft, and for something as large as an O'Neill colony you'd never notice it under normal circumstances; but if you throw the ball up, or loft it into an arching trajectory, the effect can be *very* noticeable.

(I know: it isn't *really* that way at all. To an observer watching from outside there is no such thing as "centrifugal force," and the Coriolis effect I described in the last paragraph is also a pseudo-force. What happens is that the released object tends to fly along in a straight line tangent to the circle of motion; but the effect, as far as someone inside the colony is concerned, is as I described it. I've diagrammed the situation below.)

The result is that if we did have baseball in an O'Neill colony, the batted ball would follow an abnormal trajec-

72

Figure 25

Coriolis "Force" displaces dropped or thrown object; view as
seen by observer rotating with system.

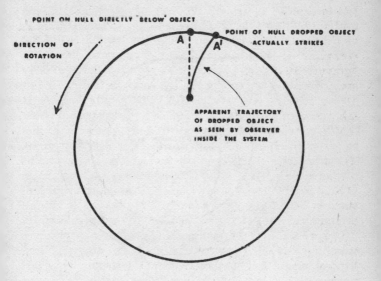

POINT ON HULL DIRECTLY "BELOW" OBJECT

POINT OF HULL DROPPED OBJECT
ACTUALLY STRIKES

DIRECTION OF
ROTATION

A

A'

APPARENT TRAJECTORY
OF DROPPED OBJECT
AS SEEN BY OBSERVER
INSIDE THE SYSTEM

Figure 26

Cause of the Coriolis Effect: true situation as seen by non-rotating observer.

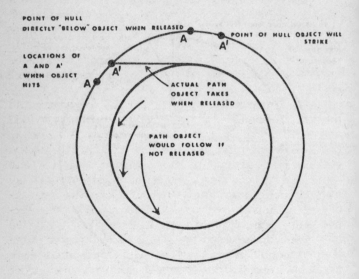

POINT OF HULL
DIRECTLY "BELOW" OBJECT WHEN RELEASED

A

A'

POINT OF HULL OBJECT WILL
STRIKE

LOCATIONS OF
A AND A'
WHEN OBJECT
HITS

A'

A

ACTUAL PATH
OBJECT TAKES
WHEN RELEASED

PATH OBJECT
WOULD FOLLOW IF
NOT RELEASED

tory. The fielders could jump fifty feet into the air in an attempt to catch it. If the ball nevertheless falls into the outfield and a player snags it, he'll have to be careful not to aim his throw at the catcher. Exactly what his point of aim should be if he wishes to get the ball to home plate will depend on where the player's standing when he makes his throw. If the axis of the field is along the axis of the cylinder it could make quite a difference whether you threw from right or left field!

* * *

Conditions in a lunar colony would be rather different. While it's only one-sixth Earth's, the gravity on the Moon is real, not artificial. Also, O'Neill colonies have to be built with a lot of open space. A Lunar base doesn't, and most models have the colony carved out of caves. It's certainly possible to roof over a large crater, and it will probably be done: but I doubt that there will be any larger surface cities.

Lunar farmers have a problem. The Sun doesn't shine all the time. During the long Lunar night there's got to be heat and light for their plants. There are a lot of schemes to provide that, from full-time artificial light to mylar-roofed craters with an opaque roof that can be put on over it (and artificial lights, of course). You certainly have to cover any transparencies (large ones, anyway) during the night cycle. If you didn't you'd lose all heat to radiation. The effective temperature of outer space is about -200° C (73° K) and heat radiates proportional to the fourth power of the temperature difference. Even here with Earth's atmosphere to catch some of that outgoing heat it's always *much* colder on a clear than a cloudy night, and in fact the Romans used the night sky to make ice cream in the Sahara. Maybe I'd better explain that.

Take one large pit, and fill it with straw. The idea is to insulate it as thoroughly as possible. Put a small container in the middle of the straw. At night you expose the pit to space. It radiates heat. In the daytime you keep it covered with more straw and on top of it all place highly polished shields or other reflective surfaces. Ice will form in a few days (provided that the night sky is clear, as it is in the desert). Enough for the Romans. Back to space.

Life on the Moon has been thoroughly described in science fiction stories, and there's no point in my doing it again. For an excellent book on the uses of the Moon, see Neil Rusczic's WHERE THE WINDS SLEEP. The lunar colony is, after all, a complex cave with lower gravity than Earth's.

Zero gravity is another story. Any long trip through space will have to be made by Hohmann transfer orbits, which use the lowest amounts of fuel, but which also take a lot of time: about a year and a half to get from Earth to Ceres, for example. Even a trip to Mars takes something over eight months.

It's possible to design ships so that they have artificial spin gravity, of course. There are some problems with that, and since many of my readers like to do their own preliminary design work, I'll give the equations here. Readers uninterested in the details can skip the next paragraphs.

Newton's First Law says that an object in inertial space wants to continue at the same velocity (that's both direction and speed) forever. It takes a *force* to make any change in velocity. Gravity serves as the force to get moving objects into an orbit, exactly as the string serves to provide a force when you whirl a weight around on the end of a rope. In both cases the object wants always to go in a straight line, which is to say it wants always to go off in a direction tangent to its circle of motion. It does *not* "fly out from the center," although the result, as seen by an observer *moving with the system*, looks that way.

Thus if you stand on a moving carousel it feels as if you're trying to fly out radially from the center, and in free space the "floor" of a centrifuge will be "down." If you let go of an object it will experience an acceleration relative to the carousel, and for those inside the system that looks very much like gravity.

The acceleration is:

(equation one)

$$a_R = w^2 R$$

where we have used w in place of the Greek letter "omega" as a kindness to typesetters. R is the radius of the rotation, and w is the rate of rotation in *radians* per second. There are 2 pi radians in a circle, so if you multiply radians per second by 360 and divide by 2 pi, you get degrees per second. Multiply the result by 60 and you have degrees per minute; divide the end result by 360 and you have revolution per minute.

(equation two)

Going the other way, $\dfrac{\text{rpm} \times 2\pi}{60}$ = radians/second.

Since force equals mass times acceleration (the most basic equation in Newtonian physics), it's easy to see that the force exerted by (and the tension on) the cord when you whirl a weight on a rope is,

(equation three)

$$F = ma = m\,w^2 R$$

where m is the mass of the whirled object. This is the centripetal force, and it's real. If the cord were suddenly cut, the object would fly away in a straight line tangential to the radius of rotation. The velocity it would have is

(equation four)

$$v_T = R\,w$$

and we're finished with the math.

* * *

Now we're ready to design a ship, and immediately we see the problem. The shorter the radius, the faster you have to spin the ship to get a given artificial gravity. Now it happens that the faster the spin, the worse the Coriolis effect. If the radius of rotation is long compared to, say, the height

77

of a man, there's no big problem, but as it gets short there can be devastating physiological effects.

It seems silly enough now that we've put men into orbit, but at one time planners seriously thought space stations and ships needed something like a full Earth gravity to keep humans alive, and we did plans for such things. If you try for a full g in a ship of small radius, the Coriolis effect is so severe that a water-hammer is set up in the circulatory system. A man could kill himself of stroke simply by turning his head rapidly in the wrong direction.

We now know that humans don't need a full gravity, and we suspect that a tenth might be enough forever. That can be arranged for a long trip if we send ships in multiples: join the ships with long cables and rotate them around each other. That's also very inefficient, of course: we have to duplicate life support systems, etc. There's less dead weight in one large ship than in many small ones.

Also the tension in the cable can get quite high, as you can find from equation three.

Maybe we don't need any gravity at all? True, the first Apollo astronauts came out much the worse for wear, and so did the first Skylab crew; but the interesting part is that the longer men stay in space, the better they adapt to it. The Skylab Four (third manned Skylab, in NASA's screwy counting system) crew came out in much better shape than did the second crew. Okay, in retrospect maybe it's not so surprising that the longer you stay in zero-g the better you adapt, but it did in fact surprise a number of space physiologists who had thought that a month of zero-g might be beyond human endurance.

* * *

A long trip in no gravity can be interesting. The accounts of the Skylab experience make for fascinating reading. They also show the need for experience in space. There were some terrible design faults in Skylab.

For instance: Skylab was the first space vehicle in which the astronauts ate at a table using spoons and forks, rather than squeezing everything from tubes and baggies. Their table was a mere pedestal that supported their food trays. There were seats, but those were seldom used: to stay in a sitting position in zero gravity requires that you bend at the waist and hold yourself bent. It puts a constant and severe strain on stomach muscles, and in fact those were the only muscles better developed when the crew landed than when they went up. The real problem, though, was the table itself.

It didn't do a very good job of holding the trays, to begin with. They tray lids were held down with what Lousma called "the most miserable latch that's ever been designed in the history of mankind or maybe before." Pogue said of the table, "I wouldn't want the people that designed that table to do anything else . . ."

Despite their attempt at normal meals, the Skylab astronauts never had much appetite. Part of that is due to less need for food: you're not working very hard in zero gravity. Also, the thinner air (kept at low pressure to avoid strain in the pressure bulkheads and such) doesn't transmit food smells very well. Everyone had head congestion, caused by pooling of body liquids in the torso and head, so nothing tasted very good anyway. However, they did eat.

With food in plastic bags (which were inside cans, which were supposed to be fitted into the trays on the table, but which often drifted loose because the cans didn't fit the trays very well) they could use spoons and forks. Eating in zero-g takes practice. You have to be careful to bring the spoon in a smooth arc from tray to mouth. Any hesitation and the food travels on in a straight line, probably into your eye.

The Skylab astronauts were almost constantly dehydrated, but never felt thirsty. The human organism is designed with a number of mechanisms to get the blood back out of the legs and up into the torso. So long as the legs are below the body those work fine; but when there's no such thing as "below," the blood gets into the torso and stays there. With all that fluid pooled in the abdominal region the thirst mechanisms don't work well, and the Skylab crews had to train themselves to take a quick drink every time they passed the water fountain. The fountain wasn't designed very well either, with metal nozzles that would have been easy to use on the ground, but which could chip teeth when not under control. The fountain buttons were so stiff that when a crewman pushed one, the button didn't go down, the crewman went up, unless he was holding onto something.

Of course it's hard to blame the designers. Until Skylab nobody had any real experience at designing living quarters for space. Apollo was a ship, and there wasn't much room to move around in it. The crew mission was to get somewhere and come back, not live in space. Gemini was worse, and Mercury was downright primitive: when we stuffed people into the Mercury capsules they were fitted in precisely, without even room to straighten arms and legs. John Glenn once said you don't ride a Mercury capsule, you wear it.

And prior to Mercury we hadn't any real experience at all. We flew transport planes in parabolic courses that might give as much as 30 seconds of almost-zero-g, and that was all we knew. I will not soon forget some of our early low-g experiments. Some genius wanted to know how a cat oriented: visual cues, or a gravity sensor? The obvious way to find out was to take a cat up in an airplane, fly the plane in a parabolic orbit, and observe the cat during the short period of zero-g.

It made sense. Maybe. It didn't make enough that anyone would authorize a large airplane for the experiment, so a camera was mounted in a small fighter (perhaps a T-bird; I forget), and the cat was carried along in the pilot's lap. A movie was made of the whole run.

The film, I fear, doesn't tell us how a cat orients. It shows the pilot frantically trying to tear the cat off his arm, and the cat just as violently resisting. Eventually the cat was broken free and let go in mid-air, where it seemed magically (teleportation? or not really zero gravity in the plane? no one knows) to move, rapidly, straight back to the pilot, claws outstretched. This time there was no tearing it loose at all. The only thing I learned from the film is that cats (or this one, anyway) don't like zero gravity, and think human beings are the obvious point of stability to cling to ...

Future dwellers in zero gravity won't have so much to worry about. The nine Skylab crewmen dictated hours and hours of notes on design improvement, this time not theory, but well founded in experience. The next space station (if we get one) should be a lot more comfortable.

And life in zero gravity, the Skylab crew tells us, is fun. Almost no one simply went from one place to another. It was impossible to resist turning somersaults, flips, ballet twirls, just for the sheer hell of it. Most of us saw the TV demonstrations: waterballs floating in air, tiny planetary systems that could be set in motion by blowing gently on them. There were other lovely experiments, and just plain play, all described beautifully in a book I recommend, Henry S. F. Cooper's A HOUSE IN SPACE (Holt, Rinehart and Winston, 1976).

* * *

Then there are the asteroids, which are different again. They have *some* gravity, but not much, making them different entirely. Things do fall, but not rapidly. On Ceres, for example, you can jump about 125 feet into the air (oops! into space) and it takes over a minute for the round trip. On very small rocks you can jump clean off, never to return. There are dangers on intermediate sizes, too, ones too large to jump from.

For example, some respectable asteroids, several kilometers in diameter, have such low gravity that if you jumped hard you'd not leave it forever, but it would take hours to go up and come back down again. You could easily run out of air.

And so forth. I've tried to describe some aspects of life in the asteroid belt in my stories "Tinker" and "Bind Your Sons to Exile," and other SF writers have written hundreds of such. It will be interesting to see how well we've done: despite all the SF stories about zero-gravity (and a number of SF fans among the engineers who designed Skylab), there have been a lot of surprises when we actually got up there.

But—no one questions that we can go. Man can live in space, and by doing so can save the Earth. We have only to want to go.

PART FIVE:

A GENERATION OF WONDER

COMMENTARY

In the past ten years we have learned more about our Earth and our solar system than did all of mankind through all our history. In the next twenty years we will again double our knowledge. This is a unique generation, a generation of wonder; for no other will ever have such an experience.

It is not that we are more intelligent than our ancestors, for we are not; in some ways we act less intelligently than they. But we have finally arrived at the equipment that lets us learn about our universe. Without deep submersibles we could never have learned about the ocean floors. Without spacecraft we could never have got a close look at other planets—and thus had data for comparison to increase our understanding of Earth itself.

The information curve has never been steeper; in ten years we have learned as much as mankind did in the previous three million. Even that time span changes like dreams. Richard Leakey, Director of the Kenya National Museum, believes he has a skull, #1407, that dates *genus homo* to three million years—and now states that he has a new find, #1805. "If 1407 bothers you, 1805 will horrify you," he told an audience at the 1974 AAAS meeting.

The computer revolution proceeds so rapidly that I own a computer more powerful than the world's best of only 10 years ago; any citizen can now have access to more computing power than was available to the most heavily funded project of the 60's.

Microcircuits let the deaf hear and the blind see, the dumb speak and the lame walk.

The science fiction of the 60's is outdated, gone, destroyed by science; although writers used the best information available at the time, the space probes have sent our old Mars and Venus into the dustbin. But if the science fiction writers have been embarrassed, what of the scientists on whose theories the better writers have relied?

"The spacecraft hang like swords of Damocles over the heads of the astronomers," Carl Sagan has said. "And on their faces you can see a strange amalgam of fear and hope as the probes approach their destinations."

Before 1980 we will have a close look at Saturn. A few years ago we knew nothing of the rings. Now we have bounced radar off them, to find that they're not dust, but chunks several meters in diameter. Within a few years Pioneer will give us an even closer look at them. Meanwhile we find there are rings around Uranus; Saturn is no longer alone in ringed splendor.

We have discovered the Van Allen Belts and the solar wind. Biologists have made giant strides toward cracking the genetic codes. Skylab has shown the way to new materials previously undreamed of. Soon the Shuttle will take a large telescope to space: a telescope capable of finding terrestrial-sized planets circling the nearer stars.

There is a serious search for extra-terrestrial intelligence, as Frank Drake and his colleagues listen for messages from Out There.

I have been privileged to watch all this. One of the great rewards of my business is attending science conferences, space launches, planetary probe encounters: of *being there* when history is made. It has been a decade of wonder— and thanks to my readers, I was there.

This chapter has been put together from a number of those conferences; it has no theme other than as a potpourri of marvels; but I think it will not be dull.

A Potpourri

Did California discover Europe? Will fusion power be used in the Light Water Reactor fission system? Whatever happened to super-heavy elements? Can we build public highways to space?

Every year the Council for the Advancement of Science Writing puts on a week-long series of briefings. The 1976 meeting was different in that the National Association of Science Writers received a generous grant from the Upjohn Company to let them bring in writers who might not otherwise attend, and the NASW for unfathomable reasons thought it would be a Good Idea to have me come down and deliver my thoughts on creative techniques in science writing. Whether my contribution to the program was much use I certainly learned more from my colleagues than I gave them; the bottom line was that for a week I heard briefings from some of the top experts in a number of fields ranging from Dr. Jonas Salk on influenza vaccination to Berkeley's Albert Ghiorso on super-heavy elements; physics, medicine, anthropology, and more. It was all fascinating; I'll see how much of it I can share.

* * *

Until very recently we thought we understood the genesis of man: African origin, long period of evolution from ape to proto-man to Neanderthal Man; then, about 32,000 years ago Cro-Magnon (or Modern) Man appeared in Europe, quickly displacing his Neanderthal cousins. Cro-Magnon man then spread across the globe and simultaneously differentiated into races. Some 11,000 years ago mongoloid (but still modern man) hunters crossed the land bridge from Asia to North America, and 500 years ago came the Europeans and Africans; and here we are.

It may not be that way at all.

Dr. C. Ranier Berger, Professor of Anthropology, Geography, and Geophysics at UCLA, reports recent archeological finds that cast doubt on the Cro-Magnon aspects of this sequence, while Dr. George Todaro, who holds the unlikely (for making anthropological discoveries) position of Chief of the Laboratory of Viral Carcinogenesis, National Cancer Institute, has evidence that mankind evolved first in Asia, not Africa. All rather disturbing, and if any of my readers are looking for a field in which to make really startling contributions, I suggest a career in anthropology. The whole subject is due for a radical and fundamental restructuring. We really don't know very much about Man's prehistory.

First, Dr. Berger and the North American Indians. You can pick up mammoth bones all over Southern California, some of them 50,000 years old and more, and anthropologists have often suspected that a few of these mammoth bones are from beasts that died an unnatural death at the hands of persons unknown; but until recently there was no real evidence of this.

Out on Santa Rosa Island, a privately owned ranch that was once connected to the mainland, one finds both mammoth bones and what can only be hearths. The bones are burnt, and were fairly obviously cooked for someone's dinner. Recent finds make that virtually certain—and radiocarbon dating shows those hearths are more than 40,000 years old. The people associated with them are mongoloid, rather definitely modern man—so, it appears, the earliest evidence of what we call Cro-Magnon Man is now found, of all places, in California, leading naturally to the question, did the New World populate the Old? Alas, though, that's not certain. What we have are some undatable chunks of human bone which seem to be of the same strata and age as the cooked mammoths; stone tools and cooking hearths definitely more than 40,000 years old; but not the two together, yet. It would be nice indeed if we could find the burial grounds, some remains of those Paleolithic hunters who slew wooly mammoths on Santa Rosa Island 40,000 years ago.

All right: modern man was in the New World well before we thought he ought to be. So what? Well, the problem is, how did he get here? The Asian land bridge across the Bering Straits opens and closes periodically with the advance and retreat of the glaciers during Ice Ages, and is the only really credible route (unless you credit the Eskimo with a far older culture and water-faring technology than anyone ever has dreamed of); which of those temporary periods brought men from Asia to North America? There have never been any Neanderthal remains found over here. None at all—and at the moment, the oldest remains of Modern Man we know of seem to be associated with California. Very interesting.

Now back to viral cancer and human evolution. No one knows whether viruses cause cancer, are caused by cancer, or prevent cancer; indeed, each of those statements is true under certain circumstances, and actually it's more confusing than that: certain viruses certainly cause cancers, but those same infectious viruses are actually generated, *created*, by healthy animals who are themselves more or less immune to that kind of viral cancer. The theory is that these animals have evolved the cancer-virus-creating mechanism as protection—a kind of self-vaccination process. That evolution takes a long time.

The worst offender is the baboon, some species of which constantly contaminate their environment with cancer virus. Fairly obviously, any animal susceptible to that form of viral cancer has got to evolve protective mechanisms; certainly a species that's immune to baboon viral cancer has a better chance of survival.

Now of the great apes, the gorilla and chimpanzee are the most closely related to Man. This is not in dispute on scientific grounds except as part of a general attack on the whole evolutionary hypothesis mounted mostly by religious authorities. (The Catholic and most Orthodox churches have long ago come to accomodation with evolution, but many Protestant sects continue to oppose the theory, and some of their spokesmen have excellent scientific credentials; I don't care to get into that discussion here.) If though you accept that Man, the gorilla, and the chimpanzee all had a common ancestor, as most evolutionary theorists do, the viral evidence becomes important: because the chimps and gorillas, alone of the Great Apes, have evolved defenses against African baboon viral cancer. Neither Man, nor the gibbons, nor the orangutans have done so.

Baboon virus is infectious to Man, New World monkeys, and Asian apes. Incidentally, the common house cat has also come to terms with baboon virus, but not totally, arguing that the cats reached Africa well before Man, but not as long ago as gorillas and chimpanzees.

Dr. Todaro's conclusion is that somewhere after Man, the Gorilla, and the Chimpanzee differentiated—say 12 million years ago—gorilla and chimpanzee ancestors made their way to Africa and stayed there. Man did not, but must have spent nearly the whole of the Pliocene Era, all that time until perhaps a million years ago, in Asia. The data, he says, "suggest that . . . the older Australopithecines found in Africa, though clearly hominids, were probably, therefore, not in the main lineage to Man, but rather, unsuccessful offshoots whose progeny have not endured to the present." (Beneviste and Todaro, "Evolution of type C viral genes," *Nature* Vol *261*:101, 13 May 1976)

Which brings us back to may earlier statement, that if you're looking for a field that needs some really new contributions, anthropology is ready for a new genius.

So much for anthropology; now for something practical like energy. There were two speakers, Dr. Robert Thresher of Oregon State University who is part of an ERDA project on wind energy, and Dr. Moshe Lubin of the University of Rochester on fusion.

There's either not much, or far too much, on wind: that is, there is no startling new information, only a very great number of studies and experiments designed to inch our way forward to a time when wind might provide as much as 5% of our national electric power; and while 5% is respectable and very much worth working on, it's not going to change the world.

The largest windmill ever built was a 1.25 megaWatt machine on Grandpa's Knob in Vermont. It was called the Smith-Putnam machine, it worked in the 40's and it was a failure: it couldn't compete economically with coal, and eventually suffered a catastrophic accident. (Windpower experts study "loss of blade accidents" the same way that nuclear engineers study loss of coolant accidents in fission plants.)

There was once a 200 kiloWatt machine working on the city island of Gentzer in Denmark; at present that mill is tethered, but the Danes are thinking of refurbishing it. What's important are the numbers: a modern electric plant generates something like 1000 megaWatts; the biggest windmill ever made was 1.25 megaWatts; and ERDA's big new experimental windmill, the Mod Zero constructed near Cleveland to study stresses and strains of putting all that much metal up in the sky, is a 100 kiloWatt device. It takes a *lot* of windmills to make significant amounts of power, which isn't to say that windmills won't be useful, particularly in remote windy places far from other fuel sources.

Fusion, on the other hand, is generating a bit more excitement. You'll recall from previous columns that fusion has its ups and downs: a few years ago, everyone thought it was the new hope of the future. Last year at the AAAS meeting you could cut the gloom with a knife. The present mood, according to Dr. Lubin (whose work at Rochester is in laser fusion) is one of controlled optimism.

First, nobody has changed their mind: fusion will not produce direct on-line power in significant amounts before the years 2010 to 2020, exactly as I've reported previously.

Second, the national energy plan still calls for about 50% of US baseload electric power to come from nuclear *fission* by 1990; and thirdly, present uranium reserves cannot sustain nuclear fission power at the rate of consumption for more than forty years.

That's the energy dilemma: we *need* nuclear power. If you think strip mines are bad now, wait until 1995 without nuclear, when there will be enormous freight trains running about the country carrying nothing but coal; coal-slurry pipelines will cross the deserts and rivers and wild places; black-lung compensation payments will be in the tens of billions of dollars a year. The precipitated flyash and other waste products will accumulate in *billions of tons*, and must be disposed of somehow, and even then millions of tons of pollutants will get into the atmosphere even with the best cleanup technology. We need nuclear power, which is to say fission power (the only kind we've got) to get to the end of the century: but the nuclear power fuels can't last very long after that.

Two ways to go. One is to make more nuclear fuel, which is to say breeders. I used to be a big enthusiast for the fast breeder, and I'm still willing to argue the case for them; after all, breeder technology, which we invented, is now in use in England, France, and the Soviet Union although we've yet to build a commercial demonstration plant. However, the breeder has its problems. Plutonium is nasty stuff. The nuclear fuel cycle has vulnerable points in it, times when terrorists might be able to get their hands on weapons-grade plutonium, or at least manage to get something that could be chemically refined into a weapon. Nobody, deep down in his heart, loves plutonium (but nobody really loves blacklung and other coal side effects, though we already put up with them).

What would be really marvelous would be a system that lets us invest in conventional Light Water Reactors (LWR's), a proven technology that we've got on the shelf, and operate them without nuclear fuel reprocessing. It turns out there may be two ways to do this.

One is the "slow breeder": Thorium, a relatively plentiful element, can be bred into U-233, which can then fuel conventional reactors. That's relatively expensive power compared to burning natural uranium, but it has the advantage of being a nearly eternal source of energy. Alas, it also requires a new technology, including mining and refining techniques, and it doesn't do anything with the truly monstrous amounts of uranium we've already mined.

There is enough U-238 around in mine tailings, stockpiles, etc., to last the world at least a thousand years. The value of the U-238 already mined in the US is one trillion dollars—a national treasure indeed if we can use it. U-238 won't fission, though, and has to be bombarded with neutrons to turn it into plutonium—and we've already discussed that. Nobody wants all that plutonium.

But suppose we could make the plutonium safe? Paradoxically, the best way to do that might be to make it more dangerous. That is, nobody in his right mind is going to try to steal spent fuel elements. "Used" fuel rods contain not only long-lived plutonium, but also various fission prod-

92

ucts, which are short-lived and thus *very* radioactive. You don't want to get close to them, and if you have the technology to work with things like that, you don't need to go steal your fissionables: that is, it takes something like a wealthy government to be able to make useful weapons out of spent fuel rods. The dangerous part of the nuclear fuel cycle comes when the plutonium has been extracted and is lying about by itself; that can be handled with only moderate care.

So now comes the point. Fusion power systems produce neutrons. (For a lot more on this subject, see "Fusion Without Ex-lax.") When neutrons interact with U-238, they turn it into plutonium which can be used to power an ordinary LWR. Suppose, suppose we took spent fuel elements, left them in the sealed rods, and inserted them into a "recharging" system? Can we do that?

According to Dr. Lubin, we can.

The current status of fusion research is summarized in Figure 27. You can see there are some problems, but we're moving toward getting useful power from fusion devices. However, as will be pointed out in "Fusion Without Ex-Lax," once you have achieved fusion you still do not have a useful power plant. What you've got is a lot of fast neutrons; they still must be caught and their energy extracted. You've still got to build turbines and generators or a big MHD (magneto-hydro-dynamics and don't worry about it) tunnel, or some other very massive and very costly system for taking neutron energy and turning it into electricity. That can be a very large problem, although it hasn't been emphasized much by fusion enthusiasts.

But—fusion makes neutrons. Neutrons are what's needed to "recharge" spent fuel elements. Spent fuel elements, whether "recharged" or not, are so dangerous that they're safe: that is, they can be shipped about in huge containers stressed to withstand hundreds of g's, and nobody is going to open one of those things. With "recharging" there is never a point in the fuel cycle where weapons-grade material exists; the Pu concentration in a "recharged" fuel element will be around 5% of the oxide (while weapons-grade is about 90% enriched metallic) so that even if a demented terrorist group stole the fuel elements they'd be useless. (Oh, sure, they'd be dangerous, but so would the equivalent weight of TNT or plastique.)

At any rate, the concept is fascinating, and provides one bit of evidence for *my* basic thesis: that we are not doomed, the Club of Rome is wrong, and mankind has a very good chance at "Survival with Style."

Now, in keeping with the title of this book, let's get far out.

In Blish's classic *Cities In Space* series one basic element was antiagathic drugs: a pharmacology that cures death by reversing the effects of aging. Now in principle such things must be possible: certainly it must be possible to take a human being at some arbitrary stage of development and stimulate continuous regeneration so the system never "wears out." "In principle" is not practice, though; nobody knows how to do this.

However, Dr. Allan Goldstein of the University of Texas Medical Branch, Galveston, may well have taken several giant steps down that road. Dr. Goldstein, with Dr. Abraham White at Albert Einstein University, some years ago began work on immunological deficiencies in humans. All our textbooks tell us that the thymus gland, that lump on the breastbone, degenerates at about age 40 to 50. The older textbooks say the function of the gland is unknown; bolder spirits even asserted that it was useless, something like a vermiform appendix. That turns out not to be the case.

Human beings have two periods of severe danger: in childhood, before the immune system develops; and in old age, when the immune system deteriorates. In both those times we are vulnerable to various cancers, infectious diseases, and auto-immune disorders. Next, let us plot the

Figure 27

THE CURRENT STATUS OF INERTIAL CONFINEMENT FUSION RESEARCH

Kind of Reaction	Power (10^{12} Watt)		Efficiency	
	Obtained	Required	Obtained	Required
Lasers				
Solid State	3–5	30–50	0.3%	0.5%
Gaseous	0.3	200	0.5%	0.5–5.0%
Charged Particles				
Electrons	0.5	400	20%	10%
Ions	—	600	10%	?
Neutrals	—	?	—	?

10^{12} Watts = 1 terraWatt or TW.

Inertial confinement: pellets of Deuterated polyethylene about 100 microns in diameter (1 micron = 1/25,000 inches) are bombarded with particles.

levels of thymosin in the blood at various ages. (Thymosin is one of the secretions of the thymus gland.)

The results are interesting: as the thymus gland vanishes, which it does until at age 40 only about 10% remains, and at age 80 it is virtually gone, the thymosin level falls, and our susceptibility to diseases of aging—those very ones that were so dangerous to us in childhood before the immune system developed—mounts rapidly.

Dr. Goldstein has used thymosin to treat children with immune system deficiencies. The results have been dramatic. Not many studies have been done: although several children were selected for this treatment, most died before the FDA gave permission for this very new drug to be used in humans.

The obvious next step is to try thymosin in persons age 40 and above, bringing the level up to what it was when they were 20 or so. That may take a while: it is estimated that it would cost $30 million in studies to get aspirin approved by the FDA even given what we already know about it; I wouldn't care to estimate what the costs of getting thymosin approved might be.

However, Dr. Goldstein has pretty well proved that the thymus gland is the "master gland" of the immune system, and that treatments with thymosin have been very useful for very young children with immunity disorders; that thymosin stimulates the development of certain cells which control phagocyte (while blood cell) cancer control activity—that is thymosin stimulates development of T-cells; T-cells somehow detect cancerous mutations and secrete a substance that brings phagocytes to the area; and the phagocytes eat up the cancer cells before they can multiply. Dr. Goldstein is emphatic in stating that thymosin is not the "magic bullet" for curing old age—but he strongly suspects that it can be useful in letting one age gracefully, without many of the pains and ailments so common in those over sixty.

He's also rather excited about all the developments in

biochemistry and immunology. We are on the threshold of a new era in medicine. Understanding the immune system will of course make transplant technology much more reliable; may provide the key to cancer, and almost certainly will help keep patients alive long enough for other cancer treatments to be effective; and may well be the means for all of us to stay alive gracefully at least to the biblical three score and ten.

I have always had the view (not original with me) that the human organism is designed to self-destruct shortly after age 40. In a tribal society we ought to have the good grace to die when our children reach child-bearing age, with a few of us hanging around to be tribal elders, but most getting out of the way. Primitive communities which don't have lots of old people have more food for young ones, and their tribes increase. Modern technology changes this; now technology may find a way to overcome the self-destruct mechanism; and I find it no surprise to discover that our immunological master gland quietly vanishes about the time we're forty years old . . .

I wonder if the FDA will ever let physicians give thymosin (which is already used in treatment of cancer patients and young children) to normal people of middle age? I think I could find a number of volunteers.

Stepping a bit further out, alas, we take a step backward. According to Dr. Albert Ghiorso of Lawrence Berkeley Laboratories, we have *not* found element 126 and the "magic island of stability." Pity.

Dr. Ghiorso is probably the discover of element 104. I say probably, because the Russians like to claim they found it first. I haven't space to review all the evidence. The upshot is that an international committee has been appointed, three Soviets, three Americans, and three neutrals. The committee has never met, but it is supposed to decide who, the Americans or the Soviets, gets to name 104. (As of 1978 it still has yet to meet.—JEP)

Meanwhile, Ghiorso has reviewed the evidence of the

97

Florida State–Oak Ridge National Laboratory collaboration of Cahill and Gentry, which had hoped to find element 126 in primordial samples (the Soviets were searching for it in very old stained glass window leads) and found it wanting. Working from the other direction—if you can't find it in nature, can you make it?—Lawrence Laboratories and the Soviets at Dubna have been bombarding ^{248}Curium$_{96}$ with ^{48}Calcium$_{20}$ in an attempt to create superheavies—and found none.

It's a great pity because I've just finished a science fiction novel whose plot depends on the discovery of superheavies; but all is not lost. It's true that we haven't found any natural superheavy elements, and best efforts haven't made any, but the search is still on and they're still theoretically possible.

Finally: what makes the Sun shine? It does, you know. Some theorists now wish it didn't. (Sure: that would kill us all, but doggone it, it sure wrecks good theories....)

Open any astrophysics or intermediate astronomy textbook, and you'll see confidently asserted a series of equations showing where the Sun gets its energy. Take four protons (hydrogen nuclei) and squeeze like mad; out come four alpha particles (helium nuclei) plus two positrons plus two neutrinos. Adding the mass/energies of the input protons and subtracting out the masses of the output discloses some mass missing: enough to generate 25 million electron Volts, and thus the Sun shines.

So, a number of years ago, theoretical astrophysicists devised an experiment which would confirm this so generally accepted theory. It wasn't supposed to be an exciting experiment; but after all, we know more about the Sun than any other star, most of our astrophysics theories are deduced from stellar observations and most of those are of the Sun, and it always helps to have confirming experiments of basic theory. Hans Bethe settled it theoretically back in 1939, but it couldn't hurt to do an experiment even if this was the best understood aspect of astrophysics.

So, out in the old Homestake mine, was installed 100,000 gallons of perchloroethylene, C_2Cl_4, and a very elaborate system for counting what happened when neutrinos struck the chlorine. (That generates argon.)

The [37]Chlorine to [37]Argon reaction expected from solar neutrinos was worked out by Dr. John Bahcall, Professor of Natural Sciences, Institute for Advanced Study about fifteen years ago. The unit is the "SNU" (pronounced "snoo"), about 10^{-36} captures per target atom per second: not very many, meaning that one needs a *lot* of [37]Cl and a long time before you expect to see anything happen.

Raymond Davis and John Evans of Brookhaven National Laboratory worked out the actual test equipment, which involves finding 15 argon atoms per month in that immense tank of cleaning fluid. They have also tested the procedure, injecting known numbers of argon atoms into the system and recovering them. To the best of everyone's knowledge that experiment ought to work, and the neutrino capture rate in the tank ought to be about 6 SNU.

The observed result: a maximum of 1.3 SNU, and possibly none at all. This is astounding. Has the Sun gone out?

Dr. Bahcall is a careful man. He wanted it clearly understood that he still believes the textbook proton-alpha reaction is the explanation for why the Sun shines. However, when pressed, he will discuss what he calls "cocktail party" theories: that is, theories that a scientist might put forth in a cocktail party, but which one has no business publishing in a serious journal.

"Unfortunately," Dr. Bahcall told us, "a lot of cocktail party theories have been published . . ."

There are three major classes of theories to explain why we have observed no solar neutrinos: those that horrify astronomers, those that horrify physicists, and those that drive both up the wall.

The astronomers like to think something happens to the neutrinos on the way here: they're produced all right, but they're a lot less stable than physicists thought they were.

After all, the only observations of neutrinos have been in paths from a few centimeters to a kilometer or so long; perhaps over longer distances they decay into something else. Most physicists don't care much for that theory.

Physicists, meanwhile, have always felt that astronomers don't really understand stars as well as they think they do. Thus, Dr. Bahcall says, the failure of the standard theory just proves to physicists that they're right in being skeptical about what astronomers say. (Not that Bahcall himself has this attitude, but it is widespread.)

The result, anyway, has been what Bahcall describes as a theoretical orgy, mostly of "cocktail party" theories. Item: the Sun has "gone out" and periodically does so, reigniting after a period of gravitational collapse. Item: there's a black hole of around 1% of the Sun's mass dead center in our star, and the Sun shines because matter falling into the hole gives off energy; there's no fusion in there at all. Item (a theory that really drives astronomers nuts): suppose all the heavy elements in the Sun are concentrated in the outer layers (for reasons no one can give); then the results would be consistent with neutrino observation.

Whatever is the explanation, there's probably a Nobel Prize in it, which may explain why the Soviets are spending enormous sums, really a *lot* of money, on solar neutrino experiments. They're scaling up the Davis experiment by a factor of 10 in a tunnel under a mountain (these things need to be down deep to keep the cosmic ray counts low enough so that the solar neutrinos won't be hidden in a fog of interactions).

There's one final possibility, strange, but not out of sight: that the Sun operates, not on the

$$4P \rightarrow 4\alpha + 2_e{}^+ + 2\nu_e + 25 \text{ meV}$$

reaction I described earlier, but through the PeP reaction: a proton plus an electron plus another proton yields deuterium (heavy hydrogen) plus *one* neutrino i.e.,

$$P + e^- + P \rightarrow {}^2D_1 + \nu_e$$

which as you can see produces just half the number of neutrinos, and they're at a lower energy level too, so that the expected SNU should be about 0.3—and that's just consistent with the observed data. (NOTE: there's no proof that we have found *any* solar neutrinos; but the likely level is in the order of 1 SNU.)

Now the astronomers will not like it if the Sun turns out to run on PeP; but their unhappiness is as nothing compared to what will result if experiment shows no solar neutrinos at all. Lower than 0.3 SNU requires something really far-out, strange, new, different, a theoretical restructuring along the lines of Einstein's work.

All of which proves we don't understand our universe quite as well as some of us think we do, and that shouldn't be any surprise. Stand by. In anthropology; immunology; energy technology; astrophysics; in these and many more fields, exciting things are happening. Like it or not, the Age of Marvels is not over, and I confidently expect that about half the things I think I know will be obsolete in five years. We don't even know what makes the Sun shine!

* * *

It's 1979 and we still don't know. I thought when I wrote the above that before it could be published in book form we'd have new data. We do not, and the unfinished story is still the best we have.

Unfortunately the same is more or less true in fusion research. President Carter has given the program something less than enthusiastic support. His predecessor cut the budget, but Carter cut it again. One wonders if the US government *wants* fusion.

In 1976 a Soviet expert named Rudikov came to the United States. The Soviets are very interested in fusion; they need the energy, to help them develop the vast regions of Siberia and Turkestan and Soviet Asia. They are also quite aware that US science and technology is generally more advanced than their own, although in some areas of fusion research they lead us. Rudikov went to great pains to get the Soviet authorities to declassify his work; he came to the US to propose a cooperative effort, and gave a briefing at which he showed the work he had been doing and the results he obtained.

The US authorities immediately *classified* his talk. They quite literally hung a blanket over the blackboard Rudikov used in his lecture! If you wonder from whom they were keeping this secret, join the club. I wonder too.

Then in November 1977 Nikolai Bassov, the Soviet laser expert and Nobel Prize winner, came to the US and at a meeting in Fort Lauderdale, Florida, presented *his* results—including the announcement that his laser fusion experiments have exceeded the Lawson Criterion by a factor of five: i.e., he has reached scientific breakeven.

Once again the US authorities classified the information. Meanwhile Carter has cut the fusion budget to the bone, and as I write this they're laying off people in the fusion labs. Carter declares war on the energy crisis and as his first marching order disbands the armored divisions; in his address to the nation he used the word "research" precisely once, and that in passing; he mentioned fusion not at all.

—JEP, Hollywood, Spring 1978

Highways to Space

Some readers will recall that in my science fiction stories I often postulate laser-launching systems: that is, a very large laser that stays on the ground provides the energy to put spacecraft into orbit. It wasn't my invention: I took the concept wholesale from a paper by A. N. Pirri and R. F. Weiss of Avco-Everett Research Center, and *they* got the concept from an earlier paper by A. R. Kantrowitz. It's not only feasible, it seems inevitable.

The concept is rather simple. Take a number of lasers, and shine the output of each into a mirror. Use the mirrors to direct all that laser energy into one big steerable mirror.

The spacecraft look normal enough, and can mass up to about a metric ton (2200 pounds). They have a bell-shaped rocket chamber at the bottom. The laser energy is directed into that chamber. It heats the air in there; the air, being heated, comes out through the nozzle—exactly as does the heated gas from a conventional rocket.

Now pulse the laser beam; about 250 times a second

seems to work. Enough air gets into the rocket chamber to provide reaction mass; the capsule rises, with the laser beam tracking it as the mirror is steered. When the capsule gets high enough so that the air is too thin to work as reaction mass, fuel from on-board tanks is pushed into the rocket chamber; the laser still provides the energy (and because it does there's no need for heavy pumps and compressors and such).

Eventually the capsule gets to space: the laser cuts off and a very small solid rocket is lit off to provide the last few pounds of thrust to put the capsule in orbit.

That's the concept, and I think I was first to use it in a science fiction story. Imagine my surprise, then when at an AAAS meeting I heard Freeman Dyson give a lecture on laser-launched systems as "highways to space."

Dyson is, of course, one of the geniuses of this culture. His Dyson spheres have been used by countless science fiction writers (Larry Niven cheerfully admits that he stole the Ringworld from Dyson). One should never be surprised that Freeman Dyson—perhaps I should rephrase that. One is *always* surprised by Freeman Dyson. It's just that you shouldn't be surprised to find you've been surprised, so to speak.

Dyson wants the US to build a laser-launching system. It is, he says, far better than the shuttle, because it will give access to space—not merely for government and big corporations, but for a *lot* of people.

Dyson envisions a time when you can buy, for about the cost of a present-day house and car, a space capsule. The people collectively own the laser-launch system, and you pay a small fee to use it. Your capsule goes into orbit. As I have proved elsewhere (in the first *Galaxy* column I ever wrote, entitled "Halfway to Anywhere," April, 1974), once you're in orbit you're halfway to anyplace in the solar system. Specifically, you're halfway to the L-5 points, if you want to go help build O'Neill colonies. You're halfway to the asteroid Belt if you'd like to try your hand at prospecting. You're halfway to Mars orbit if that's your desire.

America, Dyson points out, wasn't settled by big government projects. The Great Plains and California were settled by thousands of free people moving across the plains in their own wagons. There is absolutely no reason why space cannot be settled the same way. All that's required is access.

Dangerous? Of course. Many families will be killed. A lot of pioneers didn't survive the Oregon Trail, either. The Mormon's stirring song "Come Come Ye Saints" is explicit about it: the greatest rewards go to those who dare and whose way is hard. And if we've such a horrendous surplus of people on this planet, why is it that the same people who are so enamored of Zero-Growth also want to protect everyone from every conceivable risk? Dyson's vision is different. Perhaps his first name has something to do with it? Because he's right, you know. That kind of Highway To Space would generate more true freedom than nearly anything else we could do; and if the historians who think one of the best features of America was caused by our open frontiers, and that we've lost much of our freedom through loss of the frontier—if they're right, we can in a stroke bring back a lot of what's right with the country.

Why don't we get at it?

* * *

"Come Fly with Me"

Hot Diggity Dawg! I am so excited I can hardly stay in my seat.

Which, when you come to think of it, is a pretty strange reaction to my having been to a scientific/engineering conference, namely the "Third NASA Conference on Radiation Energy Conversion;" or is it? Before I'm through with this chapter, I hope to have you jumping out of your chairs too.

The conference title doesn't tell much; but look at the session titles, and maybe it will begin to dawn. Laser Energy Conversion; Space Solar Power; Radiation Enhanced Chemistry; Solar Pumped Lasers. No? Well, I admit that the bare titles didn't exactly turn me on, either. I did have to attend: so far as I know I published the first science fiction story in which spacecraft are launched by ground-based lasers ("High Justice," which is included in the book by the same name, available from Pocket Books if you haven't bought it yet) and one paper was to be on laser-powered spaceflight. If that wasn't enough, there were also papers on solar-power satellites, and given my interest in both energy and space that's impossible to pass up.

Those papers were excellent, and I'll tell you about them in due time; but the real kicker was presented by Kenneth Sun, a graduate student at the University of Washington, the paper itself co-authored with his professor, Dr. Abraham Hertzberg. It was called "Laser Aircraft Propulsion" and who'd have thought it would be the liveliest thing I've heard all year?

But it was. One of the problems with advocating space exploration is answering the question "Yes, but how can we make any *money* out of it? How can investments be paid off? Don't tell me about new knowledge and all that, or about pie-in-the-sky fifty years from now; what I want to know is, how can we *profit* in a reasonable time?"

And the answer to that one is hard to come up with. As I discussed in the column on space industries, we have a number of concepts that look good to make money: manufacture of exotic materials such as high-coercive-strength magnets, which is easy in zero-gravity and very costly here on Earth; biochemical research; and of course space power beamed down to Earth; but every one of those is a "maybe" because of the unknown costs and unpredictable market for the products. (Energy will have a predictable market, but the cost of space power satellites is so very large compared to the same costs for kilowatts here on Earth that you need faith to invest in SPS.)

Comes now Sun and Hertzberg to give an answer to that question.

We'll pay for space satellites by using them to power airplanes; and the savings will come in money we don't have to pay the Arabs for kerosene (otherwise known as jet fuel).

Sun and Hertzberg are quite serious—and their concept uses nothing but off-the-shelf technology. The airplane is a present-day machine; by a coincidence that hardly surprises you, the Seattle team studied a Boeing airplane. The aircraft is just barely modified: the regular engines and fuel tanks are left intact. The tail (vertical stabilizer) is twinned for reasons which will be obvious in a moment; otherwise the only modification is the addition of a new engine on top of the fuselage.

The engine looks like a regular turbofan jet engine, except that in the middle, where the combustion chamber would be if it ran on kerosene, they've placed a big box-like heat exchanger. The top of the heat exchanger is a laser target. Laser energy shines onto the target; heat goes into the system; air comes in from the front through a compressor, and is heated, just like in a regular jet engine, except that instead of getting the heat from burning kerosene it gets it from the heat exchanger; and lo, the airplane flies.

It takes off under normal kerosene power. It carries aboard enough fuel to go some 900 kilometers if it loses the laser power. It lands with its usual engines. True, there's a bit of drag from the additional engine up on top, but the reduced fuel load more than compensates. Understand, Sun and Hertzberg have made no attempt to optimize the system. They have taken an existing aircraft and (on paper —no working model yet) modified it. Everything is to be kept simple.

Kerosene at present sells for about $1.00 a gallon. It is hardly unrealistic to assume that within a few years the price will be $1.00 a liter, or about $4.00/gallon. Thus the saving is significant, given what a jet plane consumes while flying.

The effect on the upper atmosphere is all to the good, too. There's no contrail from the exhaust. No exhaust. No water vapor to form clouds. No oxides of nitrogen. Nothing but hot air.

Fine. Where are they getting the laser power?

From a satellite. A very dumb satellite. The power plant is nothing but a vast grid, something like 3 kilometers by 1 kilometer, covered with presently-available solar cells operating at present efficiencies of such. At each end of the satellite is a very large CO_2 laser with mirror for steering so that it can track the aircraft. The airplane has a small laser for "handshaking" communication with the satellite, thus aiding the satellite in tracking its target.

And that's the system. Everything, including the laser, is off-the-shelf. This being an unclassified conference there wasn't and couldn't be much discussion of the big military lasers we all know exist; but nobody seriously doubts that a laser of the proper power could be built.

The satellite is a grid of carbon-filament structure, nothing difficult to build. It can be rolled or folded up in segments and carried in the Shuttle. The grid, partly assembled on Earth, can be unrolled—if you've never seen some of the new carbon-filament stuff, you'd be amazed at how strong it can be and yet roll up like whalebone—once it's up to low earth orbit (LEO), after which it is covered with solar cells. The electricity generated by the cells is used to drive an ion engine—also off-the-shelf—which takes the satellite up to geosynchronous earth orbit (GEO).

And here's where the excitement comes in. At presently advertised prices per pound to LEO (Shuttle price) the whole project comes to less than a billion dollars. Possibly considerably less. Sun estimates about 650 million dollars. As the inventor of Pournelle's Law of Costs and Schedules ("Everything Takes Longer and Costs More") I have me doubts about that figure, but it's good ballpark range.

And—each satellite, powering only *two* aircraft per satellite, pays for itself in fuel savings alone in about thirty years.

Investment under a billion; payoff in less than thirty years; that's no longer a big government program. That's the kind of thing utilities are used to. True, utilities want more for their billion invested in a power plant than just to get the money back; they want profits; but—

It's at this point that entrepreneurial skill comes in. As G. Harry Stine says in his THIRD INDUSTRIAL REVOLUTION, somebody's going to get rich in space. There will be new billionaires as a result of space exploitation, just as there were created multi-millionaires from the aircraft industry, and from electronics. If a space project of this magnitude can pay for itself with an assured market—and it's inconceivable that the need for jet travel will decrease in coming years—then the profits can be left to "fallout" benefits. After all, anybody who thinks hard about the situation knows that once space operations become routine, there will be a number of profitable lines. The manufacture of magnets, for example, is a certain winner if you've got a space station in orbit to begin with; the doubt is whether the magnets can pay for the cost of building the station.

And the same is true for other commercial concepts. No one of them is certain to pay for the costs of the system—and nobody wants to get into a number of disparate lines of business, each needed so that the total sum of their incomes will pay for the huge investment. Business doesn't work that way. But here we have a way to pay for the station, so that the fallout discoveries are pure profit.

And at the same time we advance technology, and we reduce payments for foreign oil. If some private firm doesn't leap at this, you'd think the government would. If it appeals to you, write your Congressperson—I'm sure no one in Congress has heard of this scheme.

If all that wasn't enough, there was also Wayne Jones of Lockheed, whose paper was on laser energy relay units.

If you have your power source in LEO it goes into shadow, and you lose power. Not too good. If you put it up in GEO, your laser beam must be exceedingly tightly focused or you're losing energy. Sun and Hertzberg know this, but they've eaten the extra cost to keep the system simple.

The Lockheed group, however, has studied another way. You put your energy source in GEO all right; but then you have several relay stations, which are nothing more than fancy steerable mirrors, in Leo. This reduces the total mass to orbit and cuts system costs. Relays give other options, too.

Remember my column on Jim Baen and his electric spaceships? Well, the concept is being looked at; but the laser people think that instead of beaming power for orbit-to-orbit transfer with microwaves, which take very large antennae, why not use lasers? Thus the Earth to GEO, or Earth to Moon ship, could be powered by a *big* solar-cell array which doesn't go anywhere, with the energy sent by laser.

It's all very practical. It can work, and by that I don't mean that someday we'll be able to make it work—I mean it can work with what we have right now. Either the Sun-Hertzberg "dumb" system or the more sophisticated relay system could be started right now, carried to orbit by Shuttle payloads, and got into operation a few years after the Shuttle begins routine flights. It could be that close.

And if by now I haven't got you excited, I've failed; because just writing about this makes me want to run out and shout at people. It can happen. Right now, in our lifetimes. And think of the fallout benefits:

  learning how to assemble large structures in space;
  orbit-to-orbit transfer capabilities;
  a space "construction shack."

There are a lot of others, of course. Scientific observations. As Phil Morrison told me last year, it's insane not to study the Sun; and the best way to do that is to get out there where you can see what's going on. But ignoring all those and just looking at the business opportunities, you can see mountains of payoffs.

Now what about the drawbacks? There are some, of course. First, you're dealing with gigaWatts. The University of Washington laser-powered aircraft uses a 10 gigaWatt laser power satellite. That's a lot of Watts to play with, and a big laser. What if it misses the airplane?

Well, it won't burn down Cleveland. It might start a fire, although the damage wouldn't be one ten-thousandth of the damage sustained by a crashing airplane. Matter of fact, because the aircraft are carrying so much less fuel on takeoff—and 92% of all aircraft accidents are within 5 minutes of takeoff or landing—if one *does* crash, the damage will be enormously less; and because of the reduced takeoff weight, the chances of accidents on takeoff are much reduced.

Then too, the frequency of the laser can be carefully selected so that the power will reach the airplane—it will not fly under laser power until it is at high cruising altitude—but can't reach the ground underneath. The atmosphere is pretty good at absorbing energy. Also, don't forget that the beam sweeps across the ground at something above the speed of the airplane, so that it's moving awfully fast.

But you do have a "sungun" in orbit, and in theory I suppose some madman could modify it to do considerable damage to the countryside below. Now by considerable damage I mean something comparable to an aircraft loaded with chemical bombs, understand; I don't even mean comparable to crashing a fully-loaded 747. And if the beam is constructed to radiate in frequencies that won't hit the ground, it is no simple matter to modify that laser to another frequency. Certainly that wouldn't be done secretly by one crackpot. It would take the entire satellite crew, and some of the ground personnel as well, all in a conspiracy.

And come to that—it doesn't take all that great a conspiracy to burn down a city with gasoline. Adding the satellite hasn't given potential terrorists any capability they don't have now, although it does give them a theoretically more spectacular one. On the other hand, one suspects it would be harder to enlist the satellite crew in such a conspiracy than it would be to subvert, say, the night watchman at a dynamite factory.

Effects of adding that energy to the upper atmosphere? Unpredictable. But we already add it, along with exhaust gasses, when we fly a conventional jet plane, so it's unlikely the laser will do more harm, and there's a lot of good reason to believe it will do none at all.

So much, then, for nuclear powered aircraft. They'll work, we can afford them—recall that *one* satellite controlling two and only two aircraft will pay for itself even at $1.00/gallon kerosene—and the concurrent benefits are incalculable but great. I love the idea. I hope somebody with the wherewithal to finance them will too.

Now what of "my" laser-powered spacecraft? (I hasten to add that I did not invent the concept; the work was done at Avco-Everett by A. N. Pirri and R. F. Weiss, working on a concept which seems first to have been studied by A. R. Kantrowitz; I had their paper in front of me when I wrote HIGH JUSTICE.)

They'll work. At least that's the latest thinking. True, I seem to have made a mistake: the color of the laser is likely to be ruby red, from a CO_2 laser, rather than blue-green as I have in my stories; and the launch site is likely to be a couple of thousand feet above sea level rather than down on the flat plains of Baja where I put it. Otherwise I seem to have got it pretty straight.

(There is not far from the Baja site I used for "my" launches a high plateau; purists may imagine that the launches happen there. Baja, with its tropic location and uninhabited areas to the east, remains a good site.)

I've described the concept before, and won't go much into detail here. Basically, one takes a capsule massing in the order of a metric ton and beams a lot of laser power up its tail. The report I heard at this conference was on an on-board reaction mass concept—the capsule carries fuel, which is heated by the laser energy. In my stories I used a "ramjet" effect for the initial phases of flight, with the laser heating air until the atmosphere was thin, after which it switches over to on-board reaction mass. I asked if the ram-jet concept was rejected for cause or simply unstudied, and was told that there's no theoretical reason it won't work, it just hasn't been looked into, at least not by the speakers, D. H. Douglas-Hamilton and D. S. Reilly of AVCO Everett Research Laboratory.

So that's one that's still alive, and note the possibilities: if the laser-launching facility is constructed, the real costs to orbit have been absorbed. The flight article needn't be very expensive. Thus the prediction I sometimes make in my lectures, namely that within the lifetime of my student audiences a family can afford a space capsule which can be launched for them to live in, is still very much alive.

* * *

The conference produced a lot more. Some of the papers were on concepts that seem pretty far out even by *my* standard: how do you like windmills in space?

Imagine the following: a very large structure, looking like a "Dutch" windmill, but with blades hundreds of meters high. The solar wind—that stream of particles which constantly flows from the Sun, and which wasn't suspected until we had space probes—exerts enough pressure on the blades to turn the device. The "solar windmill" turns a generator, just as the Appropriate Technology Earthbound windmills do. Electricity is produced.

Sounds fantastic, doesn't it? And colorful, and quite pretty. Alas, it doesn't turn out to be very efficient. Note that I didn't say "impractical"; it would work, it just won't work as well as other devices would given the same investment. Still, I can see a time when solar windmills might be used as a quick-and-dirty power source kludged up by a space-faring family; perhaps a TV "Spacetrain" (something like a futuristic "Wagontrain") series will make use of the idea.

There were other papers describing far-out stuff which does look to be useful.

One of the problems with space power stations is heat.

Solar cells, and everything else for that matter, work on the principle of heat difference. The greater the difference in temperature between the front and the back sides of the solar cells, the more electricity they make.

Big lasers are not all that efficient, either, and must be cooled. On Earth that's a simple enough matter—run water through the system, and either let the water flow on downstream, or pass it through a cooling tower for evaporation; either way the heat is dumped, carried away from the system of interest. (Yes, I know: that waste heat can be a problem for those who have to live with it. True, sometimes the "waste heat" turns out to be a blessing, as it has been for the marine life off New England during the winters of '76 and '77, but it wants watching; indeed, the prospect for getting unwanted waste heat off Earth entirely ought to make the conservation-minded become space enthusiasts.)

Out in space you certainly can't simply dump the heat by sending it "downstream" nor can you afford to use evaporative cooling. Evaporation would work, all right, and splendidly, but the major factor in any space system is the cost of transporting *anything* to orbit. You simply can't afford the transport costs of coolant.

Thus space structures must be cooled by radiation. In theory this is simple enough. After all, the night sky has an effective temperature wa-aa-y down there. In practice, though, it gets more difficult. It takes surface area to dump the waste heat, and surface area means mass and large structures; and those cost a bundle to put into orbit.

Comes now John Hedgepeth, President of Astro Research Corporation of Santa Barbara. He's a no-nonsense structures man who takes the attitude of "you design the concept, I'll get it built."

The problem of waste heat has to do with density of structures; if you can have them heavy you have no problem, but since in orbit you can't, you want to get the mass per area down low.

115

Figure 28

SIZES AND MASSES

	Density	Size: projected area
Skylab	6,000 grams/meter^2	15 meters^2
Rockwell Solar Power Satellite	2,000 grams/meter^2	7,000 square meters
Halley Sailor	5 gm/m^2	800 meters^2
Echo Satellite	100 gm/m^2	20 square meters

Figure 28 shows some present-day figures. What's needed are structures with very large areas and very low weights; something in the order of 10 grams per square meter or better. That turns out to be possible, although it isn't easy.

Hedgepath presented another concept. Dust, it turns out, would be a marvelous radiating system. Dust has a *lot* of area for its weight.

Of course you can't throw dust away, just as you can't throw water away, not in space. The dust itself is valuable; any mass is. But suppose you take the dust and "throw" it across a couple of hundred meters or so; catch it, pass it through a heat exchanger to heat it up again, and throw the hot dust out to another catcher; and so forth. If you arrange the pitchers and catchers properly the net result is that the dust cools things down, but there is no motion imparted to the total system.

And that makes a dramatic picture: a tetrahedron in

space, a kilometer or so high: bright lines on dark velvet. The lines are hot metallic powder (dust sounds better, doesn't it?) madly radiating the waste heat from another large structure a couple of kilometers long and half a kilometer wide. From the big solar-cell array bright beams stab down toward Earth to power an airplane skimming along at 50,000 feet and moving at 700 miles an hour.

It's colorful—and it's also practical.

Frank Coneybear, of Arthur D. Little, presented his own summary of trends in power transmission through lasers and such. He led off by saying "You can judge my faith in technology by the fact that I have my viewgraphs but I don't have my necktie: my necktie is wherever United Airlines has sent my luggage, but I carried my viewgraphs with me." He went on to summarize what's happening in laser technology and conclude that microwaves can certainly beam power down to Earth from satellite altitudes, and the technology is well-nigh off-the-shelf. However, lasers offer some big advantages, and the whole field of laser technology is exploding. There is the possibility of solar-pumped lasers, which will take sunlight (a *lot* of it, collected by a very large mirror) as their direct input.

The conference organizer, Ken Billman, is working on another direct use of sunlight: NASA's SOLARES project, which is a system of orbiting mirrors. The large mirrors concentrate sunlight onto a particular area on Earth, extending daylight by several hours a day; the effect is to increase the growing season. There is also the possibility of modifying climate through SOLARES.

And more, none of it science fiction. These are all concepts which we could start work on next year. They are understood far better than was currently in-place space technology at the time John Kennedy set us the goal of the Moon in a decade. We could start work knowing that it's nearly certain that we'll get all our money back within 30 years, and that there's a very good chance we'll all get rich from fallout benefits.

After all, for the $50 billion a year we spend on importing oil, we ought to be able to buy any number of better systems.

Why don't we get started?

The Tools of the Trade
(And Other Scientific Matters)

This will be my fourth annual report on the State of the Sciences—that is, I've just attended the 1978 annual meeting of the AAAS, and it's time to review what's going on.

There's a lot. Before this is over you'll hear proof of immortality, see projections of history out farther than anyone I know of has ever looked, get the latest on the search for extra-terrestrial intelligence (SETI), and learn what's coming next in climate (warmer for a hundred years, Ice Age in a thousand).

* * *

Science marches on. There wasn't anything really spectacular in this year's meeting; just a confirmation of trends that we've seen before. There weren't any wild disappointments, either, except for the announcement that one of the shuttles would never go to space. Those fans who worked to get the first ship's name changed to "Enterprise" have legitimate grounds for complaint: that's the one the Administration has chosen to cancel. *Enterprise* will never go to space, and I for one can't help thinking there's a kind of grim vengeance in that.

It was mildly amusing to listen to Mr. Carter's spokesman explain why killing one of the shuttles would be good for science. Of course one had to hang on to one's sense of humor. A few of those listening couldn't, and asked just how it might help the sciences to kill off some 25% of our capability for going to space. I heard no very satisfactory answer to that. The theory is that the money saved will be available to other science projects, but none of the lucky recipients were named, and you can believe as much of the theory as you want.

Enough gloom. I really don't want to write a political column.

SETI. Surely the search for others out there is worth a note. No alien intelligence has been found, of course, although enthusiasts continue to listen. Their efforts have been hampered by the limits of their receivers: there are just a lot of possible channels on which the others might be talking, and we don't have much spare radio-telescope time anyway. Thus it would make sense to listen to a very large number of channels all at once. The only problem is that no one has ever built a one-million channel receiver.

The "how to" of such a receiver has been known for years. It wouldn't even be expensive, at least not by Federal standards—under five million dollars, probably under half that. So why has it never been done?

Because no one ever asked for the money. NASA's budget people are terrified that if they ask for a couple of megabucks for a receiver with which to listen for alien intelligence, Senator Proxmire will (a) refuse the request, (b) denounce NASA and perhaps hand them his "Golden Fleece Award," and (c) wreak terrible vengeance by chopping out several tens of millions from the NASA request.

Although this is hardly a courageous stand on NASA's part, it is, alas, rather realistic. This year, though, it is said that NASA will screw up its courage and ask for the million-channel receiver for listening to possible messages from Out There. Watch for Proxmire's reactions.

* * *

There was a lot of attention to the weather at this year's AAAS meeting. I don't suppose that comes as much of a surprise, given the terrible weather we've had lately. I wish I had good news, but in fact, the consensus of opinion among the weather and climate people is that things are likely to get worse, not better.

According to the long-range weather prediction people, what we've experienced the last couple of years is "normal"; what was abnormal, and we have no right to expect for the future, is the extraordinarily *good* weather of the past 20 to 30 years.

Now things are getting back to normal, and if that turns out not to be to our liking, well, the universe never promised us anything different. The normal climate generates highly variable weather. For reasons not clearly understood, during the 50's and 60's the weather wasn't very variable, and the climate was highly benign. For the future, if you don't like the weather, wait a few decades. It will probably change.

That turns out to have a number of consequences, of course. For the moment famine is at a minimum; there are comparatively few areas of the world in which starvation is a major contributor to the death rate. Given drastic changes in climate—and we now have good reason to expect such massive changes—there will be nothing for it: either we increase productivity, or famine stalks the land again. Not, of course, *our* land. *We* won't starve; but the universe has so arranged things that if there are to be major gains in agricultural productivity, they will almost certainly come about through intensive use of western technology transplanted to the "developing nations"—or they will not come about at all. Whether we will do the necessary development is another question.

Figure 29

A UNIFIED FIELD MODEL OF THE UNIVERSE

[Newton]

Celestial Gravity----------------------------------┐
Terrestrial Gravity--------------------------------┘

[Maxwell]

Electricity-----------------------------------┐
Magnetism------------------------------------┘
Weak Nuclear--------------------------------------┤----Super-gravity?
Strong Nuclear------------------------------------┘

Figure 30   QUARKS

Quark	Binding Particle
UP	Electron
DOWN	Electron Neutrino
STRANGE	Muon
CHARMED	Muon Neutrino
TOP* (Truth)	Tau
BOTTOM* (Beauty)	Tau Neutrino

*Evidence for existence is not conclusive.

Last year I reported that physicists were challenging the General Theory of Relativity. I may not have put it precisely that way; what I said was that top physicists were fairly sure that within the century they would have a unified field theory. *That*, however, implies the overthrow of General Relativity, because GR treats gravity as a phenomenon fundamentally different from the "forces" of nature such as electro-magnetism. In General Relativity, gravity results from distortions in the fabric of space itself; it is not really a "force" at all.

Incidentally, Einstein himself searched for a unified field theory, something to relate gravity to the other forces, and although he invented GR, he didn't believe in it.

At any event, the trend is toward unification of the fundamental forces, as shown in Figure 30.

There are also continued attempts to describe the universe in simple terms—that is, what with all the elementary particles floating around, theory has become very complex, and physicists are trying to get rid of some of the particles by showing they are made up of something else, as shown in Figure 30. Like it or not, the name given the "something else" now seems to be "Quark," and the terms "Up, Down, Strange, and Charmed" also seem destined to stay; however, some physicists such as D. Allan Bromley of Yale are resisting "Truth and Beauty" as the names of the newest candidate quarks.

The impulse toward unification theories of the universe is a very old one, of course, beginning with the Greeks and their early "atomic" models. Aldous Huxley once remarked that it made no difference whether the universe "really" conformed to the simplest explanation, or scientists were just not capable of understanding anything else; and possibly there is an impulse to simplicity rooted in the human psyche. Occam's razor need not have anything to do with the real world. Yet—there are intriguing hints that the universe may after all be built more simply than it appears.

For instance, there is that intriguing number 10^{40} which appears so often. The age of the universe, calculated in units of the time required for light to cross an atomic nucleus; the diameter of the universe in units of nuclear diameters; the ratio of the strongest (strong nuclear) to the weakest (gravity) known force. There is also the mass of the universe as measured in masses of elementary particles; that turns out to be 10^{40} squared, no small number, but there is that pesky 10^{40} again.

And of course it may be pure coincidence. "What does it all mean, Mr. Natural?"

*   *   *

Let's see. What else? You must remember, an AAAS meeting is a 5-ring circus, and every day there is far more to do and see than you can possibly get to. This year it was a bit easier, because there were more of "us"; in addition to Larry Niven and myself and Mrs. Pournelle, there were from the SF community Mr. and Mrs. Frank Herbert, Joe and Gay Haldeman, David Gerrold, Charles Sheffield, Karl Pflock, Ben and Barbara Bova, and probably some others I don't remember; this made it a bit easier to trade notes on various sessions, although it also made for longer nights. Incidentally, we found a very good Afghanistani restaurant near the Sheraton Park, where we enjoyed good food while Frank and Bev Herbert regaled us with stories of their visit to the Khyber Pass.

There was also a science fiction writers panel; it was well attended, and seemed to be enjoyed by those who came. Panelists Bova, Gerrold, and Herbert spoke of matters science fiction, probably appropriate for the audience. For myself I would have preferred that they *do* SF rather than talk about it, but I was probably alone in that wish.

*   *   *

The single most fascinating session of the AAAS meeting was a panel entitled "Prospects for life in the universe: the ultimate limits to growth." Chaired by William Gale of the Bell Telephone Labs, it featured former astronaut Brian

O'Leary, Freeman Dyson, Dr. Gale himself, Gregg Edwards of NSF, and Carl Sagan as discussant. Since neither Dyson nor Sagan can read the telephone book aloud without making it interesting, that was obviously the one panel not to miss, and none of us did. It began prosaically enough, with von Puttkamer of NASA projecting space industrialization over the next 25 years; it ended with the darndest thing I've ever seen. Understand—in a sense, these were amateurs at my business, and in fact a great deal of the panel was a bit like that, scientists playing science fiction writer with no more spectacular success than most SF writers; that is, until Freeman Dyson gave his paper.

Before Dyson we had O'Leary on asteroid mining and space colonization, themes we've dealt with in this chapter and elsewhere. Not suprisingly, O'Leary recommends use of the O'Neill "mass driver" (O'Leary is O'Neill's associate at Princeton) to move asteroids around. The mass driver is that gizmo so beloved by science fiction writers, a kind of electronic catapult to fling ships—or buckets of goo—into space. Drivers don't work from Earth, but they will from the Moon, and certainly from an asteroid.

The usual SF story uses the driver to launch ships; Mr. Heinlein used one to launch capsules in THE MOON IS A HARSH MISTRESS, and a few stories have had the drivers launching raw materials from the lunar surface. The latter is the concept O'Neill's plan for space colonies employs. O'Leary's presentation proposed using the driver to move an asteroid: power the driver with solar cells, and use chunks of the asteroid as reaction mass. I've often spoken of the concept in my lectures, but whether I heard it first from O'Neill's people I don't know. Certainly it would work.

It takes, according to O'Leary's figures, about 4000 tons of equipment to haul in an asteroid. And—one asteroid brought to high Earth orbit could provide all the materials needed to build enough Solar Power Satellites to power the entire world by the year 2000. As O'Leary was speaking I made the note "Hell, it's my lecture"; which may not strictly be true, but it's close enough. We certainly could, by the year 2000, power the world from space, and we could do it without bankrupting ourselves—and I've said all this before in other columns, and although the temptation is severe I'll leave the topic alone here.

The next lecture was by Dr. Gale of Bell Labs, and once again it was a bit like listening to my own presentation—not that Dr. Gale didn't say some things I don't, but the theme was remarkably similar to my "Survival with Style," at least at first. He began by reviewing the limits to growth on Earth itself; they are, not surprisingly, pretty severe, although not as severe as the Zero-Growth people like to postulate.

The solar system, however, provides somewhat more room. It could furnish for each of a sextillion (that's 10^{16}) people: 200 tons of hydrogen; 5 tons of iron; 5 tons of glass; 400 pounds of oxygen; 400 pounds of carbon; and 50,000 kiloWatt-hours of energy. Perhaps that's life on the cheap, and we wouldn't want the full sextillion people living here, so adjust the available wealth according to the population you like.

Dr. Gale then reviewed starship systems, none going faster than light, and not surprisingly concluded that they are quite feasible if a bit expensive. Again, so far, not a lot new; but he also pointed out that given the limits of a solar system, the impulse to build starships must be reasonably high. We could go make use of other stars.

There's only one problem with that—someone else may want the materials. In fact, if you play exponential growth games, it will be only a few thousand years before humanity will have spread far throughout the Galaxy, and may well be tearing stars apart and moving big things around. (See "That Buck Rogers Stuff" for more details.) And if *we* are pretty near that stage, why haven't others done it? Those are effects we would probably see.

Thus, perhaps we are alone in this galaxy—and according to Dr. Gale, that may be as well, because in far fewer than a million years we will want it *all* for ourselves. Meanwhile it's a race—and he does not discount the possibility of a race to another galaxy so that we can lay claim before someone else does. And do note: if you project human progress and use that as a model, it is strange that the outsiders are not here yet. (Devotees of the UFO persuasion have their own ideas on that.)

In fact, Gale notes, there is no reason why within a few tens of thousands of years humanity will not be interfering with the evolution of the universe: preventing lovely and useful matter and energy from collapsing into Black Holes where we can't get at it; making stars grow in the direction we want and need; etc. There is, Gale concludes, no limit to growth except to meet someone else as powerful as we who needs the growth materials we must have. On that note he ended.

But—of course there is a limit. The universe itself is not eternal. It can't last forever.

It can't—but perhaps we can, says Freeman Dyson.

* * *

No one could ever accuse Freeman Dyson of thinking small. His "Dyson Spheres" or "Dyson Shells," large systems for trapping the energy of the sun so that not so much is wasted, were the inspiration for Larry Niven's RINGWORLD and Shaw's ORBITSVILLE and a number of other stories. Although I didn't get the concept from him, Dyson was the first non-SF type I know of who examined the space industrialization possibilities implied by the laser-launching system I've employed in many stories. He is a modern renaissance man who thinks both broadly and deeply.

He began simply enough, by quoting from Stephen Weinberg's THE FIRST THREE MINUTES. Like many modern cosmologists, Weinberg finds that the universe is doomed, and that disturbs him. He says, "The more the universe seems comprehensible, the more it seems pointless."

Of course nearly all religions have taught that "the world" (which certainly implies "universe") will inevitably come to an end. The Last Trump will sound either from Heimdall's horn or Gabriel's. Then too, true atheist humanism has never had any answer to the feeling that it is all pointless; of course it is. (This is, incidentally, discussed brilliantly and at length in Henri de Lubac, SJ., THE DRAMA OF ATHEIST HUMANISM, Meridian 1963.) It is only a true modern who can proclaim that the universe has no purpose (in the sense that it is no more than a dance of the atoms) and at the same time bewail its pointless impermanence.

Dyson, however, did not address the theology of the universe; he stayed strictly within its physics, looking at the probable futures. The first result of this is given in Figure 31.

Now note that last number: $10^{10^{76}}$. To the best of my recollection—and that of all the others in attendance at the panel—no one has *ever* tried to project the future *that* far before; but then, one learns to expect great things from Freeman Dyson.

He was not finished, though. Granted that $10^{10^{76}}$ is a large number, still, it is finite; the universe does eventually come to an end. However—do we have to?

Now that sounds like a silly question. How can we survive the end of the universe?

First a question: is the basis of consciousness matter or structure? That is equivalent to asking whether sentient black clouds or sentient computers are possible: can we, in other words, make a one-for-one transformation of a conscious being, replacing part for part, and still have a conscious entity? And will it be the same entity?

Note that this is also equivalent to asking whether you could send a conscious being by wire—tear down the original and transmit a message that would cause an exact reconstruction at the other end.

Figure 31

THE UNIVERSE ACCORDING TO FREEMAN DYSON

	YEARS:
CLOSED UNIVERSE:	
All over:	10^{11}
OPEN UNIVERSE:	
Stars cool:	10^{11}
Galactic Cores form black holes:	10^{21}
Planetary orbits decay by tide and gravitational drag; no more planets:	10^{30}
All matter is liquid:	10^{50}
Galactic Black Holes evaporate by the Hawking quantum process:	10^{100}
All matter decays to iron:	10^{1300}
Cold Iron Stars collapse to neutron stars:	$10^{10^{76}}$

Dyson: "This chart makes a very large number of assumptions about the stability of the laws of physics. . . ."

If this is possible, then the second question: are biological entities subject to scaling? One presumes that if the basis of consciousness is structure, not the physical matter, then indeed we are subject to scaling: that we could build a one-for-one transform of ourselves, into computers, or into biological processes, that could be so scaled that the subjective lifetime is infinite.

Dyson then proceeded to demonstrate this with what I call the "Integrals of Immortality." I haven't room to reproduce them here, and I suspect that even freshman calculus would turn off my readers, but indeed they exist, and given the assumptions Dyson has indeed "proved" the possibility

129

of an infinite conscious lifetime—which is to say that for all time a properly constructed entity could accumulate new experiences and fresh information, running out of neither experiences nor time in which to enjoy them.

Couple that with the Goedel theorem of mathematics which states that there is no limit to growth: there will always be new questions which cannot be answered without new assumptions, which will themselves generate new questions which cannot be answered in *that* system: and you have something to think about.

Quite an exciting panel. As I said when it was over, I had gone to the panel expecting something interesting, but after all, these people were more or less amateurs at my business (far out speculation about the future); I certainly wasn't prepared for immortality and $10^{10^{76}}$ years!

Carl Sagan's discussion was on aliens: any moderate extrapolation of human capabilities shows that within a thousand years we are likely to be visiting other stars—either physically, or certainly with messages. Thus Fermi's question: Where Are The Others? They should have been here by now.

Sagan examined several possibilities.

First, perhaps we are very early. This Sagan rejects—our sun is not very old compared to the age of the galaxy, and almost certainly there have been others. (Assuming that there are ever going to *be* others. Continuing along that line takes us to theology, and is outside the scope of this chapter; for our purposes we assume that life can come about given the right physical conditions, and that having come about it evolves. This is not, so far as I know, at all inconsistent with religion, or at least with the Catholic religion.)

Secondly, then, perhaps high-level civilizations simply cannot exist: when they reach a certain level of capability they destroy themselves. We can certainly come up with scenarios in which that happens to *us*, and it's something to think about.

Third, perhaps advances in biology bring about immortality, and that in turn changes motivations, specifically, that immortality removes the imperative for colonization and expansion. I find this unlikely; I would think that immortals would still have an imperative for exploration at the very least; I find Larry Niven's ancient Louis Wu quite believable.

Fourth, there is the "zoo" hypothesis: we are either on exhibition, or somehow subject to non-interference regulations. Obviously this is not a new idea for science fiction readers; what's interesting is that it could be presented to a bunch of scientists without getting a laugh.

Finally, Sagan speculates, perhaps there are technologies so much beyond ours that we simply cannot imagine them: the effects of really advanced technologies are not recognized by us. If so, we have an interesting future in store—because no question about it, we are already approaching the point at which we could make others unambiguously aware of our existence.

Sagan was followed by a chap from the State Department, who said among other things that we are now learning to look far into the future, and this is "particularly due to the work of people like Carl Sagan." Now I have nothing but admiration for Sagan, and I shouldn't like to take anything from his reputation; but I think he would be among the first to say that much of his speculation is a bit old hat to science fiction writers and fans; and my feeling as I listened to the State Department chap was "They're doing it to us again!" In fact, after listening to the scientists congratulate themselves over having, finally, allowed at one of their meetings some elementary speculations of the kind that have gone on in SF convention panels for decades, I very nearly titled this column "Out in the Cold Again"; however that would be uncharitable, and I am truly grateful for the opportunity to have heard Dyson.

It isn't as if it were unknown for SF people to steal from the scientists; now we get a dose of our own medicine.

It doesn't taste very good, but what the hell.

As usual I'm running out of space before I can cover even half of what went on at the meeting; but I can't end without mentioning the dinosaurs.

Warm-blooded dinosaurs. I didn't get to the panel, but I did attend the press conference. There's something mind-boggling about the whole thing: how to understand, from a few scraps of bone and fossil, the physiology of critters that died away 60 million years ago.

They were, after all, rather successful: dinosaurs, ranging in size from about that of a modern wild turkey to beasts massing 80 tons live weight, dominated the planet for nearly 100 million years. Most of them were *big*. According to the panel chaired by Dr. Everett Olson of UCLA, more than 50% of the dinosaurs were larger than all but 2% of the mammals; 70 to 80% of the mammals alive today are smaller than the smallest dinosaur.

As Olson said this, something occurred to me: if they were so large, their major problem would be getting *rid* of heat. They are on the wrong side of the surface-to-volume relationship, just as an overweight person finds it very difficult to lose a few pounds. (There ain't no justice: it's easy for a thin person to reduce.) I mentioned that in the question period and found that this seems to be a relatively new idea—not that I thought of it first, but that the biologists concerned with the dinosaurs have only in the last couple of years looked at the beasties from that point of view.

However—Walt Disney did present the "overheat" model of dinosaur extinction in *Fantasia,* as Olson pointed out to me; and looking at the long-term climate models, there's at least a chance that this is what happened to them. They cooked in their own juices.

It's all complicated because we are not even sure where the continents were when the dinosaurs flourished: it has been that long, and they were around a very long time. If you rearrange the land masses to the best guesses of the configuration of 100 million years ago, nearly all the dinosaurs lived within 40 to 50 degrees of the equator—except that there is reported a fossil print from Spitzbergen and no one is sure that Spitzbergen was that far south even back then.

There was no real conclusion, and my apologies for taking your time with the question. It interests me, even if the best I can say about warm-blooded dinosaurs is that "nobody thinks they all were 'warm-blooded' but many respectable paleobiologists think *some* were; a few think none were." As to what killed them all off, there aren't any fewer theories now than there were a few years ago.

And at least some theorists say that the warm-blooded dinosaurs became birds, and the cold-blooded ones died or became reptiles, and what's all the problem?

* * *

The final controversy was over sociobiology, and that's important enough to warrant a full column one day. At the AAAS meeting the usual group calling itself the "Committee Against Racism" showed up to enforce its idea of scientific integrity by preventing Dr. Wilson from speaking: their brilliant idea was to shout "Wilson, you're all wet!" and pour water on him, obviously refuting his ideas. Sigh.

But for all that, it was a quiet meeting, not like the one a few years ago when the "concerned" whatever they were hit Senator—then Vice President—Hubert Humphrey right smack in the mush with a ripe tomato, or the one in San Francisco at which the Racism Committee tried to quiet Sydney Hook.

I usually like to summarize the year in science, but this year it is hard. The mood was one of optimism for the technological and scientific advances of the year, and profound gloom because of the prevailing attitude of government.

We *can* do a lot. Every year the discoveries come forth, and the promise of the future gets brighter; but for the moment at least there's the question of whether we will *do* anything about bringing forth that promise.

We have the tools. Have we the will?

133

PART SIX: THE ENERGY CRISIS

COMMENTARY

I can make some claim to having invented the term "energy crisis": in 1970 (in the ante-diluvian period before the Arab boycott) I published an article called "America's Looming Energy Crisis," and that was, to the best of my knowledge, the first use, at least in the popular press, of the phrase. At the time I said we'd best get cracking before the crunch.

We didn't, of course: and things have got worse since then.

The problem with energy is that there are so many misconceptions. "Soft" energies are said to be able to handle the problem: then you work the numbers, and find what wind and tide can really do, and discover there's no chance in that direction. But many still believe in various kinds of magic even so.

Energy is a technical field; you can't avoid getting quantitative if you want to talk about the real world, as opposed to the dream world some "concerned" ecologists live in. (When I was an undergraduate and took a course in ecology the professor sent us off to learn differential equations on the theory that calculus is the appropriate language with which to describe the effects of one process on another; nowadays it's a more than even bet that if I get a letter from someone who signs himself "ecologist" my correspondent will disclaim all knowledge of mathematics, including simple algebra.)

In these chapters I have tried to keep the math to a minimum; but I haven't been able to avoid it entirely, and I don't really apologize for that. Numbers are important. In the field of energy they are *very* important. What's the use of talking about some new process if it can't possibly produce enough energy to save us?

As for example tides. For years the "concerned" types held out tidal energy as a great hope; eventually someone worked out the numbers. It seems that if you built a dam around the entire continental USA (and wouldn't that do wonders for the environment!) and captured all the tidal energy at 100% efficiency, the resulting electricity would just about power the city of Boston.

Wind has the same difficulty: there just isn't enough of it. We all "know" about windy places, but the Department of Energy (DOE; formerly Energy Research Development Agency or ERDA) studies show that most of them are not really windy enough.

Another "soft" energy source is garbage; while fusion is sometimes held out as the hope of the world. These are examined in some detail below. My apologies for the technical details; I can only plead that sometimes the details are vastly important.

Fusion Without Ex-Lax

If a man told you "The only physics I ever took was Ex-Lax," would you put him in charge of nuclear power policy?

That's not a trick question. The founder of the California People's Lobby once said it, and he was the architect of the Nuclear Shutdown Initiatives.

Alas, nuclear power is seldom discussed rationally. There are those who fear the atom; and others who have made "fusion" an incantation, a magical formula which, when uttered, ends all rational debate about power policies.

In fact, the situation is worse than that: "fusion" is a good word, and fusion scientists are white magicians; "fission" is evil, and its supporters have made pacts with Satan to loose the evil djinn Plutonium onto this world. In these days when our representatives have shunted off primary responsibility for power policy onto the general public—or have abdicated their leadership altogether—it's important that the public deal in physics, not myth.

Now science fiction readers (and I hope, all *my* readers), unlike the Ex-Lax expert, are seldom *proud* of ignorance. Indeed, from the letters I get, and the audience response during my lectures, the opposite is true. The problem is that those who know quite a lot about the energy crisis are often acutely aware of how much they do not know; and thus are conscientious enough not to get into the debates. That's admirable, but it can be disastrous: often only the opinionated and truly ignorant have a voice.

All of which is preparatory to my arguments: I want to plug electron-beam fusion research; but I want your informed support, and since I know, again from my lecture tours, that an awful lot of people don't really understand either fission or fusion, I'm going to start with the basics. My apologies to those who find this discussion elementary. I will try, for the benefit of those whose only physics are Ex-Lax but who don't boast about that, to keep this reasonably simple.

$E = mc^2$, saith Einstein; that is, mass (m) can be converted into energy (E). To be precise, energy in ergs equals converted mass, in grams, times the square of the speed of light in centimeters per second. Light-speed (c) is 3×10^{10} cm/sec, so converting a gram of mass to energy would yield 10^{21} ergs, something like 100 kilotons, or about 100 times as much energy in all forms as each of you used last year.

The equation says nothing about how that energy comes out. A moment's thought will show you that can be important. If it all comes out as neutrinos it won't do us much good. There's no way to catch them. If it comes out as protons or electrons, we're in good shape: they're charged particles, and we can pass them through a ceramic tube with coils of wire around it to get electricity directly. (That last trick is called magneto-hydro-dynamics, (MHD) and it's a

Figure 32

POWER PLANT EFFICIENCIES
(Percentage of generated heat turned into useful electricity)

	Coal	Nuclear Fission	Deuterium Fusion
Average in present use	32.53	31	—
Best in present use	41	39	—
Expected with improvements (Theoretical: includes MHD)	55	52	66 (?)

bit more complex than it sounds; but we know how to do it. It only takes energetic charged particles.)

Unfortunately, most nuclear reactions do not produce charged particles. A great deal of nuclear energy appears as neutrons, and we can't catch them in a magnetic basket. What we can do is put something in their way. They get slowed down, or stopped, and their kinetic energy is converted into heat. We extract that heat, use it to boil water, and put the water through turbines. The turbine neither knows nor cares where the heat came from; it's all the same to it whether the heat source was burning coal, fissioning uranium, or fusing hydrogen.

The turbine system is the most efficient thing we've got for turning heat into electricity; but it's not 100% efficient and never will be, nor will anything else, including MHD. Thus let's dispel the first myth about fusion: it *may* be marginally more efficient than either fossil energy or fission, but it will still have waste heat, and will still require cooling systems. No one really knows the effective operating temperatures of fusion devices—we haven't even got anything that works in a laboratory yet—but if we assume they'll be hotter than either coal or fission, fusion systems will be somewhat more efficient than those we've got; but not all *that* much more so. Known efficiencies for fossil and fission plants, and assumed ones for fission plants, are given in Figure 32.

Fission systems work thusly: a neutron source is brought near an atom that breaks apart. Neutrons are emitted. Other atoms are broken into lighter elements and more neutrons. Some of the additional neutrons are used to break up even more atoms (chain reaction), others are allowed to bombard useless stuff like uranium-238 and turn it into useful stuff like plutonium-239, and the rest are caught for their heat energy.

Fusion goes the other way. If you squeeze hydrogen atoms together and get them hot enough, they turn into helium. The resulting helium doesn't mass quite as much as the original hydrogen: result, energy. It sounds simple, and it is. This is the reaction that powers the Sun (we think). Unfortunately, we don't know how to do it, and we may never learn. Certainly we haven't even a theoretical clue as to how to bring off stellar fusion; the temperatures and pressures involved are plain beyond us.

So, we go to the next best thing and use deuterium, which we'll call "D." There are two reactions:

$$D + D \rightarrow T + p + 3.25 \text{ MeV} \ (22{,}000 \text{ kW-hr/gram}) \text{ Eq. 1}$$
and
$$D + D \rightarrow {}^3He_2 + n + 4 \text{ MeV} \ (27{,}000 \text{ kW-hr/gram}) \text{ Eq. 2}$$

and I'd better explain what all that means before I lose someone.

First, deuterium is "heavy" hydrogen. Ordinary hydrogen atoms have one proton (p) and one electron (e), and nothing else. D has an additional neutron (n); it could be written as ${}^2H_1$ where the left superscript is the atomic weight, H is the symbol for hydrogen, and the right subscript is the atomic number.

Tritium, (T), is "superheavy" hydrogen with 2 neutrons, and could be written $^3H_1$. By the same token, $^3He_2$ is "light" helium; normal helium is $^4He_2$, and this stuff is missing a neutron.

For reasons we won't worry about here, it's convenient to measure nuclear energies in Millions of electron Volts (MeV), and I've given the textbook figures; for our purposes, though, the kiloWatt-hours per gram of material fused is more relevant. For comparison, a regular 100-Watt lightbulb will use 876 kW-hr each year if left burning; obviously a 1000-Watt heater uses 1 kW-hr each hour. A kW-hr of electric power costs between 1.2 and 5¢ to generate, and is sold to the consumer for from 2¢ to a dime (although I understand that lawsuits, strikes, and interesting administrative methods have got New Yorkers paying about 20¢/kW-hr).

The two reactions shown are equally probable. Both go on at the same time, and there's no known way to favor one over the other.

The tritium and "light" helium can themselves be made to react with more D, as follows:

$$D + T \rightarrow {}^4He_2 + n + 17.6 \text{ MeV (94,000 kW-hr/gram)Eq. 3}$$
and
$$D + {}^3He_2 \rightarrow {}^4He_2 + p + 18.3 \text{ MeV (98,000 kW-hr/gram)Eq.}$$

and I'm not giving these equations just to show off. Look at them a moment.

First, note that tritium. It's radioactive with a half-life of 12 years. We can burn up most of it with the eq. 3 reaction, but we've got to keep it from getting into the atmosphere. It's in the same situation as plutonium: a useful product that we need for power; and it should suffer the same fate as plutonium, "burning" in a nuclear reactor. Until it is "burned" though, it's one of the hazards of the power system, and there's no way to change that. It's also rare: the best way to make tritium is to bombard lithium with neutrons—which makes the lithium supply critical.

Second, note those neutrons. They must be caught if we're to extract their energy. When neutrons hit other atoms, they produce radioactive isotopes. Clever design can minimize the number of truly dangerous radioactive waste products, but can never eliminate them entirely. Thus the fusion industry will need nuclear waste-disposal, and there goes myth Number Two. True: fusion is cleaner than fission power systems; but it is not *that* much cleaner.

Third, the fuel isn't free. We can't use ordinary hydrogen; we have to extract the D from it, and that takes energy; thus, at first, fusion plants will consume more energy than they produce—just as, for the first years of their lives, fission plants haven't produced the energy it took to refine their fuels, or coal plants the energy it took to mine the coal. All will, of course, show a net energy profit after two or three years.

And finally there's the *real* problem: we don't know how to do it. The basic equations for uranium fission were known for a long time before Fermi built his "pile" in the squash court of the University of Chicago, and nature was *very* cooperative anyway: the materials needed for Fermi's experiment were cheap, easily available, and simply fabricated; the instrumentation was standard; and the control system was uncomplicated. Despite the ease with which Fermi demonstrated the feasibility of self-sustained controlled fission (it worked first time), it took twenty years to get usable power from a fission reactor.

There's no reason to believe the engineering of a practical fusion power plant will take less time; and we are not yet to the squash court. We don't *know* that we can do it at all—and we're certainly a long way from running our TV sets on electricity produced by fusing D. *I have never found an expert who believes we will have a working commercial fusion power plant in this century.* The only people who say different are not in the game—and may have very large axes to grind. "Waiting for fusion" is simply not a feasible power policy. There goes the fourth myth. Depressing, isn't it?

Since the above was written there have been some changes. First, the payments for Arab oil have gone out of sight: when your energy policy is to pay out $50 billion a year to the Arabs, you can afford almost *any* alternative.

Secondly, the Soviets have offered cooperation, sending Basov and Rudikov over to show us their results.

Thirdly, in all areas of fusion research there have been some unexpected results; some breakthroughs, other things just going better than expected.

As a result, a number of fusion experts now believe we *could* have a working commercial fusion-powered generator by 1995, possibly a shade earlier.

It doesn't matter, though: at present levels of funding we will not have a reactor before 2000. We may *never* get one: the fusion research budget was deeply cut by the Carter administration, and the level-of-effort funding now supplied does not make the experts hopeful.

Moreover, the Soviet offers of cooperation were treated rather strangely: although *they* declassified the information and brought it to us, *we* classified it, although from whom we are keeping it secret escapes me. Nor were the Soviet experts given the red-carpet treatment: Basov was not even invited to Washington. I have been told that had he said fusion would *not* work he would have received the red-carpet treatment; I have no confirmation of this, nor do I know of any reason why the Carter administration would prefer that fusion research fail.

However: if we're to save ourselves from our present policy of selling the country to the Shah of Iran and the Sheik of Araby on the installment plan, we need some "Manhattan Project" type research; and instead we are slashing the research budgets to the bone and beyond.

[JEP Spring 1978]

* * *

143

Cheer up. First, we don't *need* fusion; at least not immediately. There are other ways to power our industrial civilization; other ways to spark the Third Industrial Revolution as described by me and by Harry Stine who coined the term. We can all get rich even if controlled fusion never works. That ought to be good news. Here are some methods.

First, my favorite, the ocean-thermal system, which makes use of the temperature difference between warm surface water and cold bottom water. There's more than enough power in the Tropics to run the world, the Sun renews it constantly, and we know it will work because a working plant was built in 1928. However, my engineering friends tell me that's the hard way; and we aren't likely to have operating ocean-thermal systems before the year 2000 anyway.

So what are some other ways? Here's where I get into trouble. The easiest way of all is one we have now: good old reliable (average nuclear plant operates 9 mos. each year; fossil, 8.2 mos. each year) nuclear fission. It already works, and we've already mined enough potential fuel to last us several hundred years; moreover, there's enough U-238 in ordinary rock to operate the world high-energy economy for millenia.

Alas, that takes breeder reactors, and they're controversial. They make plutonium, and everyone knows that plutonium is "the most toxic material known to man." Ralph Nader has told us so. Of course Ralph Nader is also the man who, with fanfare, bought a manual rather than an electric typewriter "to conserve energy." My electric uses the electricity generated by about a quarter of a cup of oil each year; if everybody, all 230 million of us, had an electric typer going they'd consume a few thousand barrels of oil annually: not very much in an economy that measures oil consumption in millions of barrels a day. Maybe we ought to take a closer look at some of the other things Mr. Nader says. His heart's in the right place, but he doesn't seem to do much quantitative thinking.

Toxicity of plutonium compared to other substances is shown in Figure 33. Now we don't spread much botulin around the landscape, but we do spray crops with arsenic trioxide; in fact, we today import 10 times as much arsenic as we'd have nuclear wastes if the entire US electric system were run on nuclear fission. Now I keep telling myself that I am NOT going to write a paper in defense of nuclear fission power; that it's a political matter; but dammit, at least the public debates ought to make *sense*. It's one thing seriously and soberly to debate the advantages and disadvantages of fission over fossil fuel; but it's quite another to have such an important issue decided by mythology and demonology.

One last thing, then: nuclear wastes are radioactive. There. That ought to end the debate. Surely no one wants "nuclear pollution." If I sound sarcastic, my apologies; somebody really and truly said that to me not long ago. Meant it, too; and she was an important political party official.

So let's look at radioactivity in quantitative terms. Figure 34 shows the dose in millirems (thousandths of a rem) received by each US citizen on the average. Further, let's add a couple bits of information: of the 24,000 survivors exposed to 140 rems (140,000 mrem) at Hiroshima-Nagasaki, fewer than 200 died of cancer. The probability of developing cancer from radiation exposure is about 0.018% per rem (not mrem).

As to storage: it's true that at present most nuclear wastes are stored as liquids, hundreds of thousands of gallons of them; they leak from time to time. But liquid-storage was never intended to be anything but a temporary expedient; it is possible to take those wastes and make them part of glass blocks, and only legal, not technical, barriers have prevented that.

Glass is a very stable substance. It is practically eternal. If all the nuclear wastes accumulated from the Manhattan

Figure 33

TOXICITY OF VARIOUS SUBSTANCES

	Lethal doses per spoonful
Arsenic trioxide	50
Botulinus toxin	125,000,000
Plutonium oxide (ingested)*	0.5

*metallic Pu is never employed in reactors.

Figure 34

SOURCES OF RADIATION RECEIVED BY AVERAGE US CITIZEN (ANNUAL)

Natural Sources	millirem	Man-Made Sources	millirem
Cosmic Rays*	35	X-ray, diagnostic	103
Building materials**	35	X-ray, therapy	6
Food	25	Radio-pharmacy	2
Ground	11	Global fallout	4
Air	5	Color TV	1
Own blood (K-40)	20	Nuclear Power plants	.003
TOTAL	131	If live at boundary of nuclear power plant	5
		TOTAL	121

Total annual 252

* Varies with altitude; greatest at high altitude cities.
**Varies with material; greatest with brick and stone.

Project to present—including those resulting from weapons manufacture, which created more wastes than the power program—were solidified, the resulting block would be somewhat less than 60 feet on a side. If the entire country ran off nuclear power, all the wastes from now to the year 2020 could be contained in a block less than 100 feet on a side.

And the block could be stored under a superdome-like structure in the Mojave Desert for want of anywhere else to put it. Build a concrete dome; put in the wastes; and surround it all with a chainlink fence and the warning sign "IF YOU CLIMB THIS FENCE YOU WILL DIE." Or guard the area. Or both. Eventually we will have either a use for the wastes or a permanent storage area such as geologically stable salt mines; but certainly the Mojave would hold them for a couple of hundred years if need be.

At this point in my lectures someone generally says, loudly, "sabotage!" Nuclear plants are vulnerable to that, aren't they? Well, yes; but not very. The 4 foot steel and concrete containment is designed to take an aircraft crashing into it without rupture. Anyone stealing nuclear fuels for terror purposes has set himself a pretty suicidal task, will need vast technological resources, and won't get very many people very fast—what's the point of threatening people with an increased probability of cancer 15 years after you set off your infernal device?

Oh, sure: in theory a respectable bomb can be made from nuclear fuels, and certainly one *could* refine spent fuel to get plutonium: but to do that, in secret, requires the resources of a government, and governments don't need to steal nuclear wastes. Uranium can be bought on the open market, and you can breed weapons-grade stuff in a research reactor—as India did. Note well: of all nations that have The Bomb (including several such as Israel and South Africa which probably do but which have not announced it) *not one* used power reactors in the weapons manufacture.

147

The fact is, if you're in the terror business, kidnapping is simpler; or if you've a pash to be suicidally spectacular, try crashing a hijacked jet into the Rose Bowl on New Year's Day.

In other words, if we're going to debate power policy, let's do it right, with comparisons of risks, not scare statements. I'm willing; if you're interested, my lecture fees aren't too high, and you can reach me care of the publisher. Be prepared to pay expenses and a reasonable fee.

This has taken us a long way from fusion, hasn't it? No; because fusion, like other power systems, needs to be discussed in context. I will not recommend fusion research as a magic remedy for the world's ills.

I do recommend it, though. For all its disadvantages, fusion will be, if it works, the power system of the future.

First, it's very fuel-efficient. If we can extract all the energy from our D, we get an average of about 100,000 kW-hr/gram, which is 4 times the energy per gram obtained by fission reactors, and ten million times the energy/gram from burning fossil fuels. A few thousand metric tons of D each year could power the world.

Second, the fuel's not hard to get. There is about one atom of D for every 6,000 atoms of ordinary hydrogen. The world's D supply could come from under fifty plants with water intake valves ten to twenty feet in diameter. By world's energy supply, I mean the equivalent of some 20 billion barrels of oil a day—more than enough to let Ralph Nader have a new Selectric.

Third, we can never run out of D. There are billions of cubic kilometers of water on this Earth, and at a cubic kilometer each year we'd be able to run a long time; nor would we, as I've heard someone say, "lower the oceans"; at least not more than an atomic diameter or two.

Fourth, fusion is certainly preferable to fission: the waste-management problem is much simpler. There are fewer long-term radioactive wastes to worry about. Although we can and will (and already have) shipped tons of plutonium around without anyone being injured, the stuff *is* unpleasant, and we'd be better off without it even if fusion won't eliminate all nuclear wastes.

So how do we do it? There are two major theories on how fusion plants might work. First, remember that the goal is extremely high temperatures and pressures. You can't contain them in a material object, because either your reaction melts your container, or your container cools off the D and prevents the reaction. Thus non-material confinement systems, which means in practice Magnetic Confinement. The lion's share of all fusion research goes to that. The problems are hairy: both engineering and scientific questions remain unanswered.

The equipment is huge, complex, and (need I say it?) costly. There are blind alleys. We had stellerators, and magnetic rings, and various kinds of pinch-bottles, and every one of them failed. It isn't that they weren't worth building, understand; we have learned a lot about magneto-hydro-dynamic stability of plasmas (there's a buzz phrase for you). The current approach is a device called a tokamak, and if you need to know more about those go to your nearest library. Magnetic confinement isn't expected to produce fusion neutrons for another ten years, and few think it will, even in the laboratory, produce more power than it consumes before 1990. It got plenty of funding prior to the Carter administration.

The second approach is called Inertial Confinement. This consists of taking small pellets of D, or a D and T mixture, and zapping them with lots of energy. The zapping has to be done just right. If you're not careful, too much zapping energy gets *inside* the pellet and tends to disrupt it before fusion can take place. There are other failure mechanisms, such as lopsided zaps, and not enough zap-power.

Inertial confinement has this advantage: it is pretty certain to work. That is, the problems are more engineering than scientific. It may never work *usefully*, but it almost has to work if we get the geometry right and shoot enough power to the pellets. It has a second advantage: the equipment is much cheaper (for laboratory demonstration reactors; not necessarily for a working commercial power generating system). Thus Inertial Confinement gets about 10% of the fusion research budget.

There are two branches of Inertial Confinement: laser (photon) bombardment, and particle bombardment. Of the research money in inertial confinement, 90% or more goes to laser systems. There's a reason for that: laser bombardment just may have produced fusion already. There was some fanfare a few years ago when KMS Inc., a private company, seemed to have obtained fusion neutrons. There's still some question about just what KMS did or did not achieve, but most fusion people believe that laser bombardment will eventually work out. Right now they're looking at different kinds of lasers, and may have to invent a new one (called Brand X Laser) which will zap the pellet with enough energy, yet won't penetrate the pellet too fast.

[In November, 1977, the Soviet Nobel laureate Nikolai Basov told a conference in Fort Lauderdale that he had achieved a breakthrough: by exciting the pellets with soft x-rays prior to zapping them with a laser, he had managed to get scientific breakeven; that he had bettered the so-called Lawson Criterion by a factor of five.

There was very little about Basov's announcement in the popular press; I don't know why, because it's a rather exciting discovery. There was some doubt expressed by US research workers: did Basov get the proper temperatures? As I write this the Lawrence Radiation Laboratories at Livermore are said to be checking out Basov's results, but the whole program including exactly what Basov told us is classified, so I can report no more than that.

Laser fusion research, like all other fusion research, suffered a budget cut in the first years of the Carter administration; according to the Soviets it is a very promising line of development.   JEP Spring 1978]

* * *

Finally, there's electron-beam inertial confinement, which is only carried on at Sandia Laboratories (a non-profit corporation) in Albuquerque, New Mexico. I visited the labs recently, and came away a believer.

At the time of my visit the Sandia electron-beam fusion program was funded at about $5 million a year; nowhere near enough, in my judgment. Electron bombardment may just be *the* way to go.

They've already achieved fusion with this method. Not breakeven; but both US and Soviet experiments have definitely produced several billion fusion neutrons. The Soviets are attempting to cooperate with the US, but once again, when Rudikov came to the US to tell his results, we classified his talk. However, it's definite that he reported obtaining fusion neutrons, because that was announced in the Soviet popular press. (Evidently the Soviet Union isn't interested in keeping it a secret from whomever we are keeping in the dark.)

Electron-beam fusion has always been the least-funded of the fusion research programs, in part because it doesn't *need* as much money.

I have watched a man spend four billion dollars an hour on electricity. It made him unhappy. He wanted to spend a trillion an hour. It happened at Sandia labs: they were zapping a pellet with electrons. Of course they didn't spend four billion bucks an hour for very long: a few nano-seconds, to be exact, so the total cost of the electricity was a few dollars; but if they could have kept it up!

The Sandia equipment is impressive. It's also massive, as you'd expect, considering that they handle millions of volts. To get that they have to charge up enormous capacitance systems. Sirens wail, red lights flash, needles crawl across dials, just like in a good science fiction movie, and finally the technician puts his fingers in his ears. I didn't, in time, and the noise of a couple of mega-joules arcing into a target is not easily forgotten.

151

Even so, it's not enough. That's what costs the money: building equipment that will handle those voltages without breakdown (and breakdowns are *spectacular* around there; I didn't see one, but I saw insulators the size of a desk with inch-deep gouges burned into them). Then there's the triggering problem: that system has to discharge all its energy at the *right* time, and the right time is measured in billionths of a second. It's amazing that they can do it; but they can. I saw it done. Incidentally, the voltage amplifier systems they use are called Marx generators, which gives rise to a number of puns, political jokes, etc., and worse; when they were ready to fire someone shouted "Harpo's ready!" and another man said "Stand by to fire Groucho." Then there was the zap! and a million joules flowed for a few nano-seconds.

Mega-joules. A joule is 10^7 ergs. I was duly impressed until we got to talking after the experiment. The reason they don't have fusion yet is they just can't pump enough power through the system fast enough, but don't get discouraged. The amount of power needed isn't so very large after all. In fact, we calculated that a Sears Lifetime Battery contains about 4 mega-joules, and if we could just discharge that sucker in a couple of nano-seconds we'd have fusion.

They're building a system that they expect will do just that.

At the moment Sandia hopes to be at the squash court stage by 1983: that is, by then they hope to have proved that electron bombardment inertial confinement fusion will work and can, with a lot more skull sweat and good engineering design, eventually be part of a useful electric power system.

They're doing that on five million dollars a year. This is the system I mentioned in the beginning of this article. I think it deserves more money, because with more money they may get to the squash court before 1980. They don't promise it, but then they don't need much more money either; a doubling of their present five million annually would not only let them build more hardware faster, but also let them coordinate some theoretical work going on around the country, and bring in other scientists as consultants.

So: if you're in a mood to write your Congresscritter about energy problems, you might mention that here's a slot where, in my considered judgment, a few million bucks will do a lot of good. Please don't support this because you think it will get us to the year 2000; it's unlikely. Don't support it because you think it will eliminate our need for nasty plutonium. Do it as a present for your children; do it because we could use some decent national goals, and cheap clean fusion power is one of the better gifts the US could offer the world.

Do it because we can afford it, and it's something we ought to do.

* * *

Since I wrote that, a number of things happened. First, many of my readers did send letters to Congress. Second, President Carter cut the budget for electron-beam inertial confinement fusion research. Third, Congress restored the budget.

At present writing, the chief scientist involved with this research spends as much time in Washington trying to keep his budget as he does in the laboratory trying to make neutrons. He has never met the President, although while he was in Washington Mr. Carter gave an afternoon to Mr. Lovins of "soft energy" fame.

And finally, in Fall 1978, Princeton University announced temperatures of 60 million degrees in their magnetic confinement fusion research facility. This was a real scientific breakthrough; but the news announcement also contained a statement from the Department of Energy: "This achievement does not change the national timetable to fusion energy." Of course it does not: we have no national timetable. The expected date of useful fusion energy in the United States is never.

I don't know why, but the present administration does not seem to want fusion energy.

Can Trash Save Us?

Larry Niven used to live behind a garbage dump. Well, you wouldn't have known it, of course; although his sister used to say he lived in the "wrong part" of BelAire, it is not characteristic of that fabled community that you know the municipal sanitary land fill—read garbage dump—is only about half a mile away. On the other hand, the access road to the dump is clearly visible from the San Diego Freeway, and when I drove to Larry's house—which was pretty often back in the days when we were writing THE MOTE IN GOD'S EYE and INFERNO—I couldn't help seeing the endless stream of huge trucks trundling up into the Santa Monica Mountains with their loads of refuse.

Surely, thought I, we ought to be able to do *something* with that stuff besides bury it in what might otherwise be a very nice wilderness area.

Those who would save the world seem to do so in waves. There are fads in the eco-crusading business, and just now garbage is the big one. I say fad because the 1974

155

opus AN INDEX OF POSSIBILITIES: ENERGY AND POWER, contains not a single index reference to "trash," "garbage," "waste," or "sewage," and only one tiny reference (under "methane") to use of any of these materials; yet it was supposed to be a compendium of all the ideas on how the energy crisis might be solved. Now, however, you can't pick up a work of that sort without finding article after article on how we could be saved if only we'd *use* the energy in our sewage and garbage. Unfortunately, none of the articles I've seen give any *numbers;* they're very similar to the wistful thoughts I had while driving to Larry's house. There's so much waste and trash that surely we can get a lot of energy out of it; can't we?

I don't know. Let's see.

* * *

Let's begin by looking at the present energy situation. That turns out not to be easy as you might think; the data aren't collected together into one place, and even when you find the figures they're all mixed up. It takes a lot of patience and determination to come up with a meaningful composite. Energy analysts don't seem ever to have heard of the metric system. Everything is given in terms of British Thermal Units, or tons of coal equivalent (and not everybody has a common figure on how many Btu there are in a ton of coal) or barrels of oil per day (ditto about Btu/bbl.) or kiloWatt-hours, or whatever they're enamored of.

I've put together as good a picture as I can, and I've converted everything into ergs (I grew up with the cgs system; if you like the mks system, divide by 10^7; if you like feet and pounds, get hip). The results are given in Figures 35 and 36, which show where the US energy comes from and where it goes.

Now true: the growth projected in Figure 35 makes a number of assumptions which I haven't bothered to list (it assumes a constant real increase in GNP, for example) and the percentages in Figure 36 are going to change if the US

Figure 35

ENERGY IN THE UNITED STATES

Supply and Demand in Ergs (x 10^{26})

	1974	1977	1990
DEMAND	7.9	8.7	13.3
Domestic	7.7	8.4	12.5
Household and commercial	1.5	1.6	1.99
Industrial	2.2	2.3	2.49
Transportation	1.9	2.1	2.83
Electrical Generation	2.1	2.4	5.20
Exports	0.2	0.3	0.38
Synthetic conversion losses	—	—	0.37
SUPPLY	7.9	8.7	13.3
Domestic (81%)	6.4	6.5	11.5
Coal	1.6	1.8	3.38
Petroleum	2.2	2.0	2.89
Gas	2.2	2.1	1.74
Nuclear	0.1	0.3	2.51
Hydro-electric	0.3	0.3	0.45
Shale	—	—	0.44
Other	—	—	0.07
Imports (19%)	1.5	2.1	1.76
Petroleum	1.4	2.0	1.47
Gas	0.1	0.1	0.30

(Calculated from tables in ANNUAL REVIEW OF
ENERGY for 1976)

population continues stable; but at least we've got something to work with, a way to see just how big a problem we're facing.

Now that we know how much power we need, let's find out how much garbage we have to deal with. Actually I shouldn't use the term "garbage" with its strong negative connotation: as the authors of the energy-from-waste section in ANNUAL REVIEW OF ENERGY for 1976 point out, it's precisely that term that makes the most talented administrators avoid the municipal departments of sanitation, and makes waste-collection a job at the very lowest end of the social scale. Instead I suppose I should say "urban mineral resources" or some such. Anyway, we need to know how much of it we have to work with.

The Environmental Protection Agency (EPA), which hasn't heard of the metric system either, guesses that we produce about 3.32 pounds per capita each day (1.51 kg for the more up-to-date among us). Looking at my own household that seems about right. The figure refers to *municipal wastes*, which is everything thrown away such as trash and garbage, but does not include sewage. Since there are about 250 million of us, we get a rough figure of 415 thousand *tons* each day, or 15 million tons each year, which probably explains why the cities are running out of sanitary land fill. In metric terms we have 13.7 million tons annually, still a respectable sum. We'll stay with English system for a while because the energy figures I have for what we can get out of municipal waste are, of course, given in Btu/ton. Sigh.

Incidentally, the ANNUAL REVIEW article also gives an estimate of 250 million tons municipal waste daily, of which 175 million is domestic; and that simply can't be right. It may include sewage, which we'll deal with separately; but it's still far too large.

All right: we have this incredible pile of waste, now what can we do with it? Well, if it were dry we could burn it, with due regard to cleaning up the stack gasses to avoid

Figure 36

ENERGY IN THE UNITED STATES

WHAT DO WE USE IT FOR?

Use	Percent	Ergs ($\times 10^{25}$) (1974)
Space heating	18	14.3
Water heating	4	3.2
Cooking	1.3	1.0
Air conditioning	2.5	2.0
Refrigeration	2	1.6
Industrial	(37.2)	(29.5)
Process Steam	17	13.5
Direct Heat	11	8.7
Electric Drive	8	6.3
Electrolytic processes	1.2	1.0
Transportation	(25.2)	(20.0)
Auto	13	10.3
Truck	5	4.0
Bus	0.2	0.2
Train	1	0.8
Airplane	2	1.6
Military and other	4	3.2
Feedstock	5	4.0
Other	5	4.0
TOTAL	100	79.3

(Calculated from tables in ANNUAL REVIEW OF
ENERGY for 1976)

pollution; and in fact that's what's done with a lot of it (and sometimes without worrying about the pollution aspects, either). Few places make any effort to capture the energy from that burning waste. The stuff is merely incinerated to reduce the volume. Surely there is a significant amount of energy released, though, and if we can tap it, will we be independent of Arab sheiks and Liberian tankers?

The standard figure for the energy content of municipal waste is about 10 million Btu per ton, but there's a joker: that's per ton of *dry* weight. Unfortunately, a lot of municipal waste is anything but dry, and it takes a good bit of energy to get the water out of it before it will burn at all. Still, let's assume we've dried it, somehow, and it's all ours.

We cannot yet burn it in steam boilers. The stuff consists of all kinds of things: discarded metal beds; tin cans; old Six Million Dollar Man toys; food scraps; dead animals; discarded vacuum cleaners; coffee grounds; and you name it. It must be pulverized and sorted, and that takes energy. It also takes either a very high or a very low technology: that is, one way to sort it is by human labor, but we'd probably have to increase the size of the army before we could put the unemployed to work doing *that;* thus we have to build highly sophisticated equipment, with magnets, grates, air-stream sorters, and the like, and those cost money, and municipalities raise money primarily through property taxes, and home-owners are ready to revolt already; but let's assume all those problems solved, and we've done the sorting. How much energy can we get from our rubbish?

Comes now the math: tons times energy/tons times a mess of constants I won't bother to give. The results are interesting: 1.6×10^{25} ergs, or 4.4×10^{11} kiloWatt-hours. In other words, if we captured *all* the energy from our rubbish we could produce about 2% of the energy we used in 1974. Significant, yes. Important, perhaps. But it won't save us from Arab oil and sinking tankers.

Alas, things are worse than that. No industrial process is

100% efficient, and electrical generation is no exception. With present technology the best we've been able to do at burning rubbish is 27% efficiency, which means that if all our municipal waste were burned in the best boilers we know how to build, we'd get 1.2×10^{11} kW-hrs, or, coincidentally, some 2% of the electricity generated in 1974— and we have not counted in costs, the energy needed to process and dry, or indeed much of anything.

We could go another way. If we can persuade industries to build alongside the rubbish-disposal system, or take the rubbish to their plants, so that we can use the steam directly without turning it into electricity, we get 66% of the energy value, and that's a respectable 7% of all the process steam used in 1974; well worth trying for, if it doesn't cost too much.

So. What are the costs of all this? The ANNUAL REVIEW gives some figures, which I have recalculated to give Figure 37. (I recalculated because theirs were based on plants processing 275 tons unsorted refuse per day, an awkward figure at best.) They still don't mean very much; is this a low or a high cost? Well, one figure readily available is the capital cost per installed kiloWatt of electrical power. That ranges from around $500 for a coal-fired plant to over $1000 for some kinds of nuclear.

Assuming 27% efficiency for a refuse-burning electrical plant, we find the capital cost per kW is $1103: much higher than other kinds of plant costs, which explains why electrical utilities aren't terribly interested. For $1100 a kW they can buy a nuclear *breeder* plant, whose operating costs will be lower than the value of the fuel produced. I suspect that my figures for nuclear power costs are a bit low; they're based on research done a couple of years ago, and the court and environmental impact statement costs of nuclear power are now about as high as the costs of the hardware; but even so, the refuse-plant generator is *expensive*.

Or is it? After all, the cities have to get rid of the refuse *somehow.* If we subtract off the costs of sanitary land-fill, and a number of the other expenses of disposing of that growing mound of trash that gives mayors nightmares, our electrical plants begin to make sense after all: but only if we look at cities as a total system, and city budgets aren't prepared that way. Believe me, I know: I've been Executive Assistant to the Mayor of Los Angeles. The Department of Water and Power would scream bloody blue blazes if told they had to spend that kind of money; while many of DWP's senior administrators, people you can't do without, would go job-hunting if told their lordly department was to be combined with Sanitation. Moreover, the City Council would impeach anyone suggesting the kind of capital fundraising (and consequent increase in property taxes) a large-scale electricity-from-rubbish project would require.

Still, if we were starting over; if we could look at cities as total systems, rather than as a series of independent departments; it would make a great deal of sense to get rid of our refuse and extract the energy out of it at the same time. We could *not* run the city on its own garbage, nor is garbage a particularly efficient way to get electricity; but since you have to get rid of the stuff *anyway,* you might as well take out what you can, and the total system costs probably justify the initial capital expense.

Whether in this age, given what we've already spent, it makes much sense is not so obvious: you have to look at each city independently. I suspect that LA could sell the land set aside for sanitary land-fill (another department heard from: although Recreation and Parks knows it can't keep that land forever, right now they're not turning loose) for enough to build some good-sized plants; but I haven't done the numbers. In places where land values are not so high as here, it's more dubious.

Let's leave rubbish for a bit and get down in the sewers. If garbage can't provide more than a fraction of our energy needs, can sewage save us?

Figure 37

THE COSTS OF GETTING POWER FROM WASTE

System Used	Capital Cost	Annual Operating Cost
Pyrolysis	$30,900	$6,152
Steam generation, processed refuse	26,900	5,047
Steam generation, unprocessed refuse	25,000	4,665
Electrical generation, processed refuse	36,363	6,458
Fuel preparation	10,909	3,363

All figures in constant 1978 dollars per ton (2000 lbs) of unsorted refuse per day. Source: ANNUAL REVIEW OF ENERGY, Vol 1, 1976; Annual Reviews, Inc., Palo Alto, California.

* * *

To begin with, sewage is valuable. That should be obvious: it is only the very wealthy western nation that can afford to throw away such a valuable resource. Indeed, we spend a lot of money and energy merely to throw it away, after which we burn a lot of valuable coal and oil to generate electricity in order to fix nitrogen to make fertilizer—fertilizer that's not as good as the high-nitrogen sewage we pollute rivers with.

Secondly, there's a long technological history of using sewage as a fuel source: begin with the Indian peasant, who uses buffalo chips to cook his food, and proceed to modern methane generators.

Finally, even if we don't use sewage as fertilizer—and thus by-pass the long series of inefficiencies involved in the generation and transmission of power, fixation of nitrogen, etc., culminating in commercial fertilizer—the stuff has a high energy content.

In other words, using sewage as an energy source has a lot going for it. There are several ways to go.

First, as agricultural nutrient, allowing the crop to be the actual energy-storage system. This is widely done in the Orient, with detrimental side effects: the honey-bucket is a pretty certain means of spreading epidemics. Surely we can do better than that.

Second, as methane source: shovel sewage into a tank, let ferment in the absence of air, and out comes methane. Methane is also known as natural gas, and is rather valuable; indeed, natural gas is the most critically short item in our energy budget. A pound of dry sewage solid will produce about 3 to 5 cubic feet of methane, which sells at $3.53 per 100 cubic meters. Now my source book, with a straight face, gives both those figures in the same paragraph. I'll translate: a ton of sewage produces some 225 cubic meters of methane, and the gas is worth about $80.00. The methane production pretty well sterilizes the residual, which can then be used as high-nitrogen fertilizer. (But we will have to use some of the methane as fuel for drying it before we can sell it.)

Finally, we can produce the methane and then burn the residual. It has been calculated that doing that will let us build a system that operates at zero profit provided that we charge about $5.50 a ton dump fee: that is, those who wish to dispose of sewage must pay us.

Why is this? We're getting valuable fuel out of the system; surely we need not charge a dump fee? Ah, but the plant itself costs a lot of money, and that money isn't free. IF the plant already existed, it would make a profit; but the "profit" is lower than current interest rates, and thus the dump fee. Another of the little points usually forgotten by those who are faddishly out to save the world.

On the other hand, technology improves, and one commercial utility, Southern California Edison, was at one time trying to find ways to use sewage as fuel: they contracted to take the entire sewage output of one of the smaller southern Calif. cities free and get rid of it by burning it in their boilers. They didn't quite know how to make that work, but they were doing the research, when along came the Public Utilities Commission to tell them to stop. It seems that wasn't a justifiable use of the rate-payers' money. Utilities shouldn't engage in fuel research, they should generate power. Thus we were all protected by our government, and now a government agency will have to do the necessary research if sewage is to be burned at a profit.

Still, just how much energy could we get from this source? Well, each of us produces about .25 pounds (yeah, I know, but all the other numbers I could find were in the English system and I give up) of *dry* solid each day. That's 11 million tons a year; if it all went into methane generation systems we'd get 90 billion cubic feet of methane annually, and at 994.7 Btu per cubic foot that's 9.6×10^{23} ergs or 0.19% (2 tenths of a percent) of the 1974 energy budget. Sigh. It's unlikely to save us, isn't it? However, don't despair. We can also add the animal wastes, which amount

165

to some 25 million tons a year, and get up to 2000 billion cubic feet of methane, 2.1 x 10^{25} ergs, or 2.6% of all energy used in 1974. Better than that, it's just about 10% of the energy we obtained from natural gas in 1974—i.e., we could cut natural gas consumption by 10% a year. It's not the Earth, but it's something.

Of course it's also expensive, and technologically some time away. What I've given are some maximum theoretical figures, not what we could do tomorrow morning if we set our minds to it.

* * *

Let's see: 2% from human and animal sewage; another 2% from municipal refuse; add in another percentage point just in case: and we've come up with a grand total of 5% of the 1974 energy budget, provided that we can use ALL of the energy from our sewage and garbage and other trash. Since we cannot possibly capture 100% of that energy, I leave it to you to guess at the actual efficiencies; and I think you now see why engineers, as opposed to faddists, don't think we can run the United States on its own garbage heaps.

That doesn't mean the energies in our waste aren't worth recovering. It's particularly true in the case of animal wastes: methane generators aren't very sensitive to scale. Once you get up to a couple of tons a day, larger methane cookers don't cost much less per pound processed. Thus quite small operations could feasibly build them; and, like garbage, both animal and human wastes *must* be disposed of anyway, and dairies ought to be encouraged (through higher sewage fees and the like) to catch that energy and feed it into the national pipeline. Once again, though, each case must be examined individually: distance to pipeline; availability of water (methane cookers take lots of water); and other such factors enter here.

But having done it, we haven't saved ourselves.

* * *

If we got 5% of the 1974 energy budget from sewage and garbage (which we can't, because 5% comes only with 100% efficiency), we'd still have to depend on Arab oil. What else can we do?

Not too long ago there was in vogue a scheme to grow crops for alcohol and run much of our transport system on that. It was even worked out quite elaborately: take sewage and transport it (means not specified) out to marginal farmlands; take the urban poor off the welfare rolls and give them small holdings of that marginal land; and let them grow alcohol crops, thus relieving unemployment, getting people out of the cities, and solving the energy crisis —as well as dealing with sewage.

It sounds good. It sounds marvelous. Why don't we do it?

Well, alas, there are some problems. For one, many of those urban poor have just come to the city from marginal farms on which they couldn't make a living. It might require a larger police force and army than we have to make them go back. But leave that. Let's assume we have the labor force or can get it.

What crops produce the most alcohol? You already know that: corn (maize, as they call it everywhere but in the USA); wheat; barley; in other words, cash crops. If they grow well they'll be grown: gone are the days of our big farm surpluses. They are also all crops which respond best to agro-business techniques: there are *enormous* economies of scale. It doesn't cost anywhere near a thousand times as much to keep 40,000 acres in cultivation as it does to keep 40 going.

Then there are the costs of collection. The alcohol must be transported and distributed: first, though, the crops themselves must be carried around to the fermenting vats (since the costs of providing each farm, or small community of farms, with its own generation facility would be colossal); then the resulting alcohol must be piped to where it will be used.

All this is in competition with food crops. True: the United States probably could, at hideous capital costs, grow enough alcohol to run its transport system, and thus be free or nearly so of imported petroleum. We could even afford to do it. It would have one monstrous side effect, though: as we are now one of the few food-exporting countries of this world, we would condemn a lot of people to starvation. I suspect that the very ones who now clamor for our energy-independence achieved through croplands growing alcohol would be the first to denounce such callous behavior.

But: it is in precisely this area that research can be made to pay off. Right now growing plants are very inefficient things. They don't really convert much (1% or so) of the sunlight falling on them into useful energy. Over the millenia man has selectively bred some plants (such as maize, wheat, barley, etc.) to do a much better job than most; now molecular biology may allow us to double that efficiency in a few years.

The potential is enormous. If we had 3% efficient crops and grew them continuously, then some 2% of the US land area would provide fuel for all our electrical generators: 25% of all the energy we use, and this without fusion or plutonium or coal or oil. Well worth continuing research. A year or so ago I discussed the need for a national "Manhattan Project" in plant efficiencies. It's still needed. The development of more efficient agricultural plants would do more toward solving the energy problem than all the solar towers and windmills we'll ever build.

* * *

Finally, what does it all mean? I'm tempted to give Mr. Natural's answer, but I won't.

What it means is this: there aren't any single answers to our energy problem. No magic solutions. No fads. No mysterious "them" who prevent us from solving all our problems with a few tricks.

What it takes is high technology and a lot of research:

neither of which looks too healthy right now, given our tax laws. It takes capital accumulation, incentives, a chopping away of bureacratic red tape. I've covered a lot of that in previous essays.

I don't want to leave this before making one point that glared at me out of the research I did for this article. Were I energy czar of the US, I know what I'd do (I think—after all, I've only taken a first cut).

Our transport system uses a full quarter of all energy, and 80% of our petroleum; it uses all of the imported oil. The efficiency of the US transport system is 8%, based on a figure of 0.16 kW-hr (6×10^{12} ergs) per ton-mile. Electrical vehicles can at least *double* that efficiency, even taking into account losses in transmission over wires, charging batteries, and all the rest. If we also include the costs of pollution control, the advantage of electric vehicles looms even greater.

At the moment there's no incentive to develop electric cars and trucks and the like: there's not enough electricity. If tomorrow morning all our cars were magically converted into electric vehicles, we'd be paralyzed.

We have non-polluting means of generating electricity: nuclear power plants. No. Modern fission plants are not the long-term solution to the world's energy problem, nor are they entirely safe (although compared to the health costs of coal they come off rather well). They produce plutonium which has a number of problems associated with it. But they'd get us past any current energy crisis, and if we had enough electricity we would develop—finish, really, because we've just about invented it—the technology to convert from internal combustion to electric transport; that in itself would save more energy than we could get from burning all our garbage and sewage—and it would move us a long way toward energy independence.

Meanwhile, for long-term I'd finance a *lot* of research into more efficient energy-fixing crops; I'd put development money into ocean-thermal energy (useful because a good bit of the research goes into low-heat turbine systems which can be used as bottoming cycles on existing power plants even if they never got to sea); I'd change the tax laws to give *lower* property taxes to those who insulate their houses and install solar heaters (rather than the ruinous fines called "increased property valuation" that the city now levies on you for insulating), and I'd put a bit more into fusion. Finally, I'd change the tax laws to let all kinds of private enterprises do research in the field of energy conservation and effciencies, even if that meant lower tax revenue and thus laying off some bureaucrats.

But that's a dream world, no more realistic than the dream of those who'd run the US of A on its own garbage.

The Moral Equivalent of War

There are times when I am certain I have lost my senses. I hear put forward, by supposedly sane and rational people, propositions so mindless that I doubt my sanity. Could I be *that* wrong?

One such experience happened while listening to the President's energy message to the people. Here was the President of the United States; a man who presumably can obtain the best advice from the world's most intelligent and informed people; and what I heard came out as nonsense.

After further study it still seems nonsense.

The best summary I've heard is quoted in ACCESS TO ENERGY, an excellent newsletter published by Dr. Petr Beckmann of the University of Colorado ($9.00 year, Box 2298, Boulder, Colo. 80306, payment must accompany subscription order; highly recommended). In Beckmann's latest there is a quote from an independent oil producer: "I find it; I develop it; I operate it; I take all the risks; I get $2.50/bbl; the government gets $8.25/bbl and the price to the consumer goes up."

That's an energy policy?

Now, were these enormous new taxes—the greatest tax increase in US peacetime history—to be applied to development of new energy resources, we would be involved in a sensible debate: should energy research be directed by the government, or left to private industry? One might rationally take either side of the issue, and the disagreement would be "legitimate"; I would not have this mind-boggled sensation.

But—in the President's speech for the better part of an hour, research was mentioned precisely *once,* and then only in passing. Fusion was mentioned not at all. Neither were Space Power Satellites. There was plenty talk of "windfall profits" to the oil companies but very little of why many major oil companies *love* the new policy: it eliminates all hope of competition. There was nothing about alternate fuel sources, reclamation of sewage, or agricultural research; nothing about geopressurized domes (of which more later).

Out in the fusion research laboratories they're laying off scientists as I write this.

President Carter has declared war on the energy crisis, and his first marching order was to disband the armored divisions.

* * *

So what might be a sensible energy policy? Understand, I don't claim to be the world's greatest expert on the subject; perhaps it's possible that Carter and Company can refute what I'm saying; but I do claim that what I propose makes sense, and on the evidence mine appears a better policy than the one we seem doomed to adopt; and I have heard no refutation, not even a discussion, from the "experts."

I say this up front because recently I received an amazing letter from a professor in a state college. He had, it seems, been given a copy of one of my columns, and he wrote to tell me that he intended to use me as "an example of those who hold the view that since technology got us into our ecological dilemma, it can get us out of it." Then followed a rather imperious demand for my qualifications as an expert. I gather that creeping credentialism abounds at the professor's institution.

He signed himself as an "ecologist." That's just as well, since he can hardly call himself a logician. I cannot think where I have ever said anything as amazing as *that!* I don't accept that "technology got us into our ecological dilemma," and I certainly don't accept his syllogism. I tried to tell him in my reply that I am willing to discuss specific problems and specific technologies, but that evidently did not appeal to him, for he never answered. Perhaps I do not have the right credentials to engage in correspondence with an "ecologist."

The first job in forming an energy policy is to look at time spans and constraints. It makes no sense to put together a policy which insures a crunch a few years downstream.

Democracies have no very creditable record of planning for the future. Aristocracies and monarchies have sometimes worried about the next generation, simply because it's likely that the children of the ruling class will have to live with the problems created by present governors. One might think this would apply to democracies as well, but so far that's been rare.

Yet: energy policy affects the future in an all-too-real sense. Unless we are willing to face massive cutbacks in our real standard of living, we must prepare for the future; and by "we" I mean *us*, those my age and those just coming of voting age, and by "the future" I mean *our* future.

It's a well-known fact that there was a "baby boom" in the late Forties and early Fifties. As a result we have a work force that's fairly large compared to the total population.

This has resulted in relatively high unemployment among the young unskilled; but it has also enabled the work force to support everyone else. The Social Security program was never an actual insurance system; it always frankly depended on requiring the young workers to support the aged and retired—and even with the large numbers entering the work force over the past few years, Social Security is bankrupt.

What happens when the "baby-boom" children, those born between 1945 and 1955, retire? Who will support them? You see, the fertility rate in the US is quite low, below replacement; and for the next twenty years at least there *cannot* be much increase in the size of the work force. Even were we to have a new "baby boom" the effects on the work force would not be seen for nearly twenty years—and there are some fallout detriments to a sudden increase in population. One suspects that a campaign to increase the number of children per US family would not be an optimum solution.

But if the work force stays constant or decreases, and the number of retired greatly increases, what is the result? We've seen it already: people don't retire, they just get poor. Even ownership of a home does little to cushion the blow: property taxes are far higher than the mortgage payments, and Social Security and retirement income generally can't meet the tax bill and leave anything left over.

So: either the productivity per worker increases, or we have a hefty decline in real income for an increasingly large part of the population—and precisely that portion of the population which has time for political activities, and has been around long enough to have some idea of how the political system works.

What will the result of that be? There are a number of possibilities, none very pleasant. Probably the least unpleasant would be a backlash against "ecologists" and "concerned" people, the scrapping of conservation programs, and a crash program to increase productivity at

174

any expense.

The alternative to waiting for the crunch is to plan for increased worker productivity: and that means to expand the energy supply. The productivity of the work force has always been dependent on the availability and price of energy.

Meanwhile, across the world there is a rising tide of demand: what, when I was in college, the professors called "the revolution of rising expectations." Some of those foreign beggars are actually demanding enough to eat! In many "developing" countries there is an actual expectation of development.

I suggest that it is much harder to raise real *per capita* income from $100 a year to $150 than to raise it from $5000 to $7500; that if it is to be done, it is likely to be done only with technology; and that takes energy. (It also makes it likely that the Western nations will grow even wealthier with respect to the poor ones; but see Matthew and the parable of the laborers in the vineyard on that. I should have thought it takes a twisted logic indeed to prefer an income of $100 yearly while the West stays put to having 50% more while the West does the same.)

Well, all right; we'll have to have energy; but can't conservation do it? Or windmills? Or tide? Or magic?

No one seriously thinks so, and few say it except some "ecology" publications more dedicated to ideology than fact.

But can't we go back to the land? Conserve?

How? The productive land is in production; sending any large part of the population "back to the land" would simply reduce the amount of food available. There is land that is not economic to farm; but to suggest that we "retire" our poor and aged onto inadequate farmland requires a callous disregard for human values almost beyond belief. Subsistence farming on poor land is appealing only to hopeless romantics who haven't tried it. There may be a few who like that sort of thing; but politically it's an impossible solution, requiring armies and police and the abrogation of democracy.

No. To provide for the non-working among us, both unemployed and aged, we need higher productivity; and that requires energy.

For the near term—say between now and the year 2000 —there are only two certain sources of energy in the quantities we will need.

Coal and nuclear—and perhaps, just perhaps, a third, one just discovered which does not yet figure into anybody's energy estimates.

Coal has problems. For each 1000 megaWatts of coal-fired electrical plant, there is created each *second:* 600 lbs. of carbon dioxide; 10 pounds of sulfur dioxide; 30 pounds of bottom and fly ash; and a lot of other stuff.

If you prefer annual figures, that's nine million tons of CO_2, 157,000 tons of SO_2, and 474,000 tons of ashes; not to mention a million or so tons of sludges generated in the scrubbers. There is also put into the atmosphere considerably more radioactivity (in the form of radium and radon not removed by scrubbers) than a comparable nuclear plant releases.

Disposing of the wastes from coal-fired plants is not a simple problem. Meanwhile, by the year 2000, we will be ripping from the ground about nine billion tons of coal each year. That coal must be shipped about the country, and the number of miners and railroad workers killed each year will not be zero. The ashes and other wastes must also be shipped, and disposed of, and that's hardly a trivial activity either.

On purely humanitarian grounds, coal is inferior to nuclear power: the number of people killed per kiloWatt is about 100 times greater for coal than nuclear, when you examine the entire power cycle from mines to waste disposal; while the ratio with respect to members of the general public killed cannot be estimated, because no member of the public has ever been killed (or injured) by a nuclear power plant. As I write this, ironically, there lies on my desk an article about a bursting dam killing 31 people.

But we do have coal, and we can survive through its use; and if we get on with the job we can live with the result, hard cheese though it may be for miners with black lung (my wife's father died of silicosis). I would have thought, though, that those truly concerned for the environment would prefer the nuclear option.

Yet it's very hard to find any kind of rational discussion of nuclear energy.

Item: a science fiction writer friend, a lady I respect, called on me the other day. She wanted a dramatic incident in a story. How, she asked, might a character cause a nuclear explosion—not a large one, just a little one—at the San Onofre nuclear plant?

I told her to have her character carry an atom bomb into the plant. It's the only way I know of. Beware of the guards, and recall that San Onofre is on a US Marine base. Where they get the atom bomb I don't know.

But—isn't there another way? Another lady, a radio broadcaster specializing in "conservation" and "environment" and "concern" and the like was present at the discussion and was certain that San Onofre could be made to go up like a bomb. Just cut the cooling water supply.

She wasn't very interested in hearing of the actual engineering details of the plant—feedwater supply which could be used for emergency cooling; the emergency cooling system itself; etc. Nor was my SF writer friend, who went away disappointed and muttering about how she could fake it, since the public wouldn't know about all the safety precautions.

Item: A book published by the Reader's Digest and widely touted: WE ALMOST LOST DETROIT. Never mind that we didn't almost lose anything, at Detroit (where a couple of fuel elements of the Fermi research reactor melted, the safety devices worked as they should have, and everything shut down with neither disaster nor danger of one) or at

Brown's Ferry (where a twit using a candle to search for leaks set insulation afire and caused the plant to be shut down in an orderly manner).

Item: a review of WE ALMOST LOST DETROIT published in the *New York Times Book Review*. Written by a staff lawyer for the American Civil Liberties Union, the review states "They knew what the public did not—a mistake could trigger a nuclear explosion." Do I need to repeat it? If a horde of terrorists had taken posession of Fermi reactor and used hammers and hacksaws they could not have triggered a nuclear explosion.

Item: the last time I had a kind word about nuclear power, there came in the mail from a windmill experimenter a letter that opened with obscenities. When I wrote him to ask what contribution that made to the discussion, he said that I as a writer ought to understand that this was the only way he could express what he felt. He enclosed some more obscenities, presumably to deliver himself of more of his feelings; and did not seem to understand that I didn't invent the wind figures, nor are his "feelings" particularly important to the universe. I hope he sells lots of windmills, but he can't change the energy picture much by doing it.

(Incidentally, why he thought an old soldier would be shocked by anatomical obscenities is beyond me. Perhaps he thought my wife would open the mail—she sometimes does—and that he'd impress her? Unlikely. Despite my misgivings, Mrs. Pournelle teaches in a juvenile detention facility, and I suspect she could, given enough provocation, shock the windmill designer out of his socks.)

Item: Jack Anderson, in a dazzling display of journalistic integrity, says in a recent column: "The clouds originate from six mammoth, cylindrical cooling towers that rise from the banks of the Ohio River like idols to the gods of energy. Superheated [sic] vapors from the nuclear works below form the clouds which appear so white and innocent. But they hang over Shippingport like a pall. Beneath them is a dying town contaminated by ... deadly ir-radiated mist."

Would you believe he's talking about Ohio River water trickled down through cellotex? At the bottom of the tower there is a heat exchanger which connects with the con-denser at the low-temp side of the turbines. No connection with the reactor at all, and the towers would be there whether Shippingport was fired by coal, oil, natural gas, or uranium; but you'd never know that from Anderson, who is terrified of fluffy white clouds composed of—water vapor and nothing else.

Item: *Time* magazine's recent report on "The atom's global garbage," which states baldly that there is no tech-nology for disposing of nuclear wastes. Evidently the edi-tors of *Time* do not even read *Scientific American,* for in the June 1977 issue of *SA* there is a very complete article on the disposal of reactor wastes.

Look, can we once and for all dispose of the idiotic view that there is no place to put nuclear wastes? The technolo-gy is already proven: they can be reduced to solids im-bedded in glass. The total volume from the invention of the first reactor to the most optimistic (in my view) construc-tion program of reactors extending to the year 2000 is a cube under 100 feet on a side; in fact, if the world ran off nuclear reactors exclusively, then in 50,000 years we would have enough wastes to cover about one square mile to a depth of six feet.

There are a lot of square miles of desert in this world; and at the lowest level of technology imaginable those wastes can be stored in concrete structures in the Mojave, where they are completely recoverable if needed—and they just might be.

In the early days of this century, oil companies distilled off only the higher-grade volatiles from crude. The result-ing sludge was a mess, and it was expensive to get rid of. One oil company executive ordered the company's director of research to think of something better to do with crude sludge than simply storing it.

The result was the petrochemical industry: plastics, "coal-tar derivatives," and such. Think about that the next time someone mentions "nuclear wastes."

And as for plutonium, it's more valuable than gold; why should it be stored anywhere? The value of already-mined uranium in this land is something like a *trillion* dollars, given that we go ahead with the breeder program—Except that we won't. However, the Soviets, French, Germans, British, and Japanese are already doing so. Perhaps we will export uranium? Instead of using it to fuel our own reactors, we can sell uranium and buy Arab oil. Marvelous.

Now I don't mean to imply that there is no such thing as intelligent opposition to nuclear power; but there is very little of that. The above is a far better illustration of the nuclear debate.

But, perhaps we will have to do without nuclear power, not on rational grounds, but because people are afraid? Not if you ask the people. The anti-nuclear forces have yet to win a major referendum—but there are fewer and fewer nuclear plants ordered, and our nuclear reactor industry is liquidating itself, because it takes 63—*sixty-three*—separate permits to construct a nuclear plant, and very few companies can afford the delays and the legal fees required. As a means of subsidizing lawyers the present nuclear regulation system is well designed—but is there not perhaps a cheaper method of rewarding legal diligence? It would probably be cheaper to give each law-school graduate a guaranteed salary of $50,000 a year on the condition that he (or she) not practice law.

Incidentally, the total output of *Time* magazine for 8 months takes up more space than would all our nuclear wastes from 1944 to the year 2000. (Figure courtesy of Petr Beckmann.)

Obviously, then, my "ideal" energy policy would remove a number of the constraints surrounding nuclear power plants. I would *not* relax the safety regulations, nor would I leave plant site location to "experts" without discussion; but surely the number of permits can be reduced to four or five, and the time required to get a permit cut from five-plus years to one year.

I would also do the same for coal; there is no reason why our fuels decisions should be made on the basis of regulatory difficulty and red tape, instead of economics. Utility companies, both public and privately owned, have plenty of talent for deciding what kinds of plants they ought to build; why not let them employ it? But at the moment the Department of Energy will have a budget of $10 billion, and very little of that will go into any kind of meaningful research.

* * *

Obviously a meaningful energy policy must do more than streamline the permit system. Coal and nuclear power can get us to the year 2000, and it's hard to see anything else that can; but no one genuinely loves either as a power source. Both have drawbacks, and I'm as aware of them as anyone. I would not care to leave my grandchildren the same problems I face.

What, then, should we do for the intermediate and long terms?

Well, first, you make certain there will be *something* that works, which is why I like nuclear power; it's a proven technology with, in my judgment, fewer problems than coal. But having insured there will be power, you look for better systems.

I think few would argue: one excellent power source is natural gas. Natural gas is a wonder fuel: it's clean, it is easily and economically transported, the distribution system (pipelines) already exists; it can be employed in a conservation strategy, that is, by decentralizing power generation so that on-site plants can provide both steam and electricity for major industries and compact bedroom communities. The only problem with gas is that we're running out of it faster than anything else (and, of course, it does produce CO_2 and add to the Earth's heat burden; more on that later).

There may be a source of gas so large as to be nearly incredible. The estimated conventional reserves of natural gas in US fields is 6 billion cubic meters; and we'll run out far too soon, unless we find and develop more.

But that's conventional fields. There are now known to exist "geopressure zones"; these are large pockets of water, at very high temperature and pressure, saturated with gas. I quote from the 1977 ANNUAL REVIEW OF ENERGY: "In the northern part of the Gulf of Mexico (onshore and off-shore), the only area in the world where detailed studies of geopressure zones have been undertaken, an area of 375,000 km^2 is believed to contain a large belt of geopressure zones, some going down to a depth of 16,000 meters (16 kilometers). The large quantity of gas in this belt is estimated at 1300 trillion cubic meters (roughly 1800 billion metric tons coal equivalent, comparable to entire US coal resources); other estimates, such as the one of Dorfman at 160 trillion cubic meters, are more modest, but still impressive.

"Throughout the world, many geopressure zones of the depositional or the techtonic-occurrence types have been identified, but no estimates of their methane potential have been made. If appropriately tapped, the methane resources can provide, in addition to the natural gas dissolved, enormous amounts of mechanical and geothermal energies from the pressures in excess of 500 kg/cm^2 and temperatures above 200°C at depths of 6000-7000 meters."

There are also gas hydrates, known as "frozen natural gas," discovered by Soviet petroleum geologists in the last few years, and also investigated by the US exploration ship *Glomar Challenger*. The Soviets estimate—are you ready for this?—that one million trillion, repeat, 10^6 trillion cubic meters of methane can be found in frozen reserves throughout the world's oceans.

The ANNUAL REVIEW adds, "commercial recovery of gas from submarine hydrate is probably a very difficult task." This is why I do not put hydrates down as a near-term insurance fuel source; it is risky technology.

The same is true, but to a far lesser extent, for the geopressure zones. Drilling to 6 kilometers is difficult but has been done; extracting the dissolved gas from the pressurized water is again tricky, but off-the-shelf technology. Within a few years we could have plenty of natural gas.

If someone goes after it. The simplest way to develop this resource would be to decontrol the price of natural gas; our fuel bills would be higher (but no higher than they'll be after Carter's energy taxes!) and plenty of developers, motivated by good old reliable greed, will break their arses trying to get natural gas to sell.

Another way, one less preferable in my judgment but defensible, would be to have a government development program—which would in practice mean contracting with private firms to do the work, because the expertise exists outside ERDA, not in it.

What we're doing, though, is ignoring the whole situation in favor of taxes and a complicated income-redistribution scheme; and *that*, in my view, is mindless.

Perhaps Carter doesn't know about geopressure zones? But surely *someone* does. So why is this whole technology resource ignored in the President's war on energy resources?

* * *

For the long term we cannot continue to rely on fossil fuels. There are prefectly legitimate conservation reasons—such as CO_2 buildup, and the overall heat balance of Earth —for looking to other energy sources, and probably the best way to have those taken into account is to plan *now* while someone still cares. Given our present war on energy, the day will soon dawn when no one will give a damn about conservation, and we will go to a crash program to strip-mine coal, dig up pressurized gas, build nuclear plants, and to hell with the consequences. At the moment few outside the "snow belt" during the Great Freeze of '76 have felt a real energy pinch; we are wealthy enough to give some thought for "the environment."

The poor—and that certainly includes those rich by world standards but poor compared to their own past—generally do not care about long-term consequences.

If we have a long-term goal of eliminating fossil fuels, there are only two ways to go: nuclear fusion, and some form of solar power.

Fusion has received short shrift from Carter exactly at the time when the most scientific progress has been made. The program moved ahead, they have made neutrons in reactors, and they are ready to move to breakeven, the point at which the experiment produces more power than it consumes—only they need more equipment, and they need to keep their staffs together, and Carter's budget has provision for neither.

Obviously my "ideal" energy program restores the fusion budget at least to what President Ford recommended.

There are three ways to go with solar; Earth-based, space-based, and agricultural. The latter is a kind of Earth-based solar power accumulator, of course, but uses a different kind of expertise; biologists and agronomists rather than engineers and physicists.

Plants typically store up about 1% of the sunlight energy that falls on them. A billion or so dollars doled out over the next few years could, in the judgment of people who have some right to an opinion, at least double that. I have insufficient data on which to judge, but I do point out that the risk is low—not very much money involved, compared to the welfare budget, and if done skillfully, for a few dollars more an agricultural research program could be a part of a jobs program. And the payoff is very high. I quote

184

Jonathan Swift: "Whoever could make two ears of corn, or two blades of grass, to grow where only one grew before, would deserve better of mankind than the whole race of politicians."

Some forms of Earth-based solar research get a good bit of money; but the technological risks are quite high, except in the small backyard "appropriate technology" applications. I have nothing against those—I shall probably install a form of solar heating system for my hot-water and office heaters—but they will not save us, and they are rather expensive. As a form of conservation they are excellent, but conservation is not the answer.

Still, Earth-based solar is one program with a long-term payoff that is treated about as it deserves to be. We shall see whether, when the payoff finally comes near, the "ecologists" and "concerned scientists" will not try to halt actual installations: after all, the solar constant is 1 kiloWatt per square meter, and to get 1000 megaWatts one must cover *at least* a million square meters; probably a lot more, and this will go into the "fragile ecology" of the desert.

If the desert is too valuable to take up a few thousand square meters as nuclear waste storage, I wonder why it can be covered with little blue cells; but perhaps.

Finally, there is space-based solar power, a concept whose time seems to have come—except that it receives no mention in the President's declaration of the moral equivalent of war, nor has NASA got what anyone would consider adequate funding. I have discussed SPS systems in other chapters. For my energy program I would give NASA an additional $2 billion for booster development, thus insuring access to space; and fund at a few tens of millions feasibility studies of power satellite systems. Incidentally, the United States gives the World Bank $2 billion a year, of which $225 million goes directly for the salaries of McNamara and 374 other top executives; and if *that* $225 million is well spent, surely some development of Solar

Power Satellites would be even more worthwhile?

I haven't even mentioned other schemes, such as the Ocean Thermal System I've described before. I haven't budgeted for garbage and trash (their potential is not great, about 5% at best, but then no one seriously believes windmills can contribute much more than 5% of the energy requirement, and windmills are getting a lot of money). I haven't got far out and talked of mining the Moon, or building big space colonies.

I haven't spoken of airboats—you know, big plastic structures with sides about 500 meters high; you pump out the air and they will literally float on air; get them up in the jet stream and mount windmills on them, and send power down the tether cable. *That is* far out; although not impossible, and surely better than poverty or war?

But, curiously, nothing of this emerges from the President's battle plan. Despite the rhetoric about the obscene profits of the big oil companies, the President's moral equivalent of war has the effect of giving the big internationals a monopoly on our most vital need.

So on whom have we declared war?

CONCLUSION

Some Futures

"We shall nobly save or meanly lose the last best hope of Earth." Lincoln was talking about an entirely different conflict when he said that; but it is a statement that applies to this generation in a way that was never true of Lincoln's. No one today seriously believes that human chattel slavery would have survived into the present era no matter what the Union did in 1860; but I do seriously believe that a generation a hundred years from now might curse our memory.

With increasing frequency I am asked to lecture at various colleges and universities. My message is generally the same: that we don't have to die. Western Civilization need not be finished. This generation can, if it will, make advances at least as significant as the control of fire, the discovery of the wheel, yea, the invention of agriculture. We have only to make the decision, and a few sacrifices. The technology exists.

When I am done I find a curious and almost universal

response. First, the audience is somewhat overwhelmed, which doesn't surprise me because I have developed a lot of data and, bluntly, I'm pretty good at presenting it. Next, I find the message welcome, and few in the audience want to argue. Again not surprising: why would anyone, particularly the young, want to believe anything else? But third, what was once a surprise but happens so frequently it no longer is: "Where have you been? Why has no one told us this? All we hear is that Earth is polluted, technology can't save us and is evil to boot, we're running out of resources, there's Only One Earth..."

Nor does that response come only from students. At a major southwestern research institution I got the same response when I spoke to the technical staff. Presumably each of the engineers had opinions not too different from mine: but the prevailing climate of opinion led each to believe the others thought we were doomed. At the universities the students find the faculty members, if they comment on the future at all, crying doom.

So I shuttle back and forth across the country, to this institution and that, trying desperately to convince America's youth that they have a future. But do they?

Because of course I cannot tell what the future *will* be. I know what it *can* be: a world of plenty, a world of "Survival with Style"; a world in which the United States is wealthy but is not merely an island of wealth in a vast sea of misery; a world in which everyone has more than enough to eat, and, if they want it, a standard of living at least as good as that we enjoyed in the 50's; but that's only what *can* be.

It need not be that way at all. Our grandchildren may curse our memories. There are times when I am convinced that my world can never be.

But what could happen to us? We have the technology. There is no "energy crisis" in any meaningful sense—that is, we know how to produce the energy we need to sustain our high-technology society until such time as we can develop eternal sources.

We know how to get to space, and we could, if we had to, begin right *now* the development of capabilities for mining the Moon and the asteroid belt. The world already grows more than enough food to support its population (although the distribution system is terrible and insects, rodents, bacilli, fungi, and other pests eat more of our crops than ever we do, particularly in the developing nations). We *can* survive, and with style; why might we not?

To begin with, there are the final two horsemen. War is hardly impossible. True, there are signs that many rational planners in the Soviet Union realize that their own development—yea, and survival—depends on *not* conquering the West; that the West is more valuable as a trading partner than ever it would be as part of an empire managed as badly as the Soviet Union now is. True, but not decisive. There are dinosaurs in the Soviet Union, real communists whose moral position is intolerable if ever they abandon chiliastic Marxism. For an analogy : could an Inquisition priest ever have admitted even the possibility that his religion was not true? Could he have lived with himself if ever he did?

And the Soviet Union continues to build weapons long after any discoverable need. Recall the theory? US weapons development stimulated the Soviets; once we called a halt, and they achieved parity, they would see the wastefulness of it all—after all, weapons cost them far more than us (in terms of respective Gross National Products), and they need the resources for development far more than we do—and they would cease the arms race.

So we stopped, and they achieved parity, and they achieved superiority, and they seem headed for supremacy, and they halt not, neither do they slow; indeed, their rate of arms procurement tends rather to increase. No, war is no impossibility.

Then there are the fears of cogent men like Robert Vacca, whose book THE COMING DARK AGE cannot be ignored.

189

Vacca points out the increasing complexity of our civilization, its increasing dependence on centralized planning and control, the interdependence of all parts on each other, the far-reaching consequences of seemingly trivial errors—recall the power failure in the Northeast caused by one generator going and kicking out all the others? If the margins get thin enough, and Vacca believes they will, collapse of our civilization could be much quicker, and much more thorough, than might be supposed.

After all, no country is more than three meals from bread riots; and rioters have been known to act as if they believed the best way to feed themselves is to burn the bakeries. Urban firestorms are hardly impossible, and the water supply and fire fighting systems are vulnerable. You might or might not be surprised to know just how easily such systems could be knocked out, by accident or by design.

Imagine the colossal traffic jams if the traffic signals ceased working. Couple that with snowfall and ice. Barges frozen in mid-stream, unable to supply coal-powered electric plants; coal yards frozen solid; insufficient electricity to operate the pipelines, thus cutting off oil and gas; railroads not working; people freezing in the dark; trucks not working (it takes electricity to get the gasoline out of our environmentally-protected tanks in filling stations); goods not moving—but you need not imagine it, because it happened to some of you, briefly, and on a smaller scale. Fortunately the nuclear power plants continued to operate, and the Great Freeze of '76 was essentially local; but it takes no great imagination to envision a much widerspread catastrophe, and to couple it with deliberate action by, say, the authors of THE ANARCHISTS' COOKBOOK, to see how easily the nation could be crippled.

Temporarily: for now. We have vast resources, surpluses, and a residuum of collective loyalty and humanitarianism. Neither of these conditions need prevail. Taxes can end both, and there are signs they are working to that goal as I write this.

But I don't imagine "the collapse," the "knockout blow" that Vacca foresees, as happening this year or next. So far we have a great deal of survival-surplus in our system, and it would take no miracles to insure against the knockout; but the trend is in the other direction.

Consider. One of the most influential books of the past few years is E. F. Schumacher's SMALL IS BEAUTIFUL: ECONOMICS AS IF PEOPLE MATTERED. The "appropriate technology" movement has gained many adherents. After all, as Joe Coates (Office of Technology Assessment) said, "Who can be in favor of inappropriate technology?"

At the Denver meeting of the AAAS in 1977 the appropriate technology movement had a seminar and an exhibit. I attended both in the hopes of learning something useful. Instead I saw an interminable series of slides showing true ugliness as if it were beautiful. Photos of privies dominated: not only those $3000 Swedish gizmos that more or less automatically compost the stuff right in your own home, but also old-fashioned ODT's of the kind my wife and I experienced in our childhood; the kind with the crescent cut in the door and a Sears catalogue handy in case you run out of corncobs. "You only have to fork the stuff over about every two weeks," we were told. "Of course you can run into problems with city departments of health."

To which my reaction was that I sincerely hoped if any of *my* neighbors install a privy the Department of Health will give them not merely problems, but citations.

There was more. Bathtubs made of wine vats. A speaker who told how Appropriate Technology changes your head: when the wind comes up at 2 AM and the batteries are all charged up, and you've got work to do, why, you get up and do it. Don't waste that wind energy, because the windmill can't power things to your convenience: it is you who must adapt.

And make no mistake: appropriate technology is not merely for the developing nations alone (if at all); it's for *us*. So just what is it? According to the fact sheet prepared by the National Center for Appropriate Technology, the characteristics are "(1) small scale, (2) decentralized, (3) simple to understand and operate, (4) ecologically sound, and (5) labor intensive."

So. Leave out the first four points, and come to the last: labor intensive. That is not a mere necessary evil. It is the heart of the AP movement. Given the choice they'll take hard labor over machinery every time. I call to evidence their exhibit: a bicycle seat with pedals attached to a chain that ran a wheat grinder. You are to sit and knead bread with the hands while pumping away on the bicycle to grind the wheat with your own muscle power. There was also a film strip showing how the bicycle seat system could be attached to plows (dragging the plow through the dirt) or water pumps, etc., etc.

Now as an advance over the mortar and pestle, a leg-powered crank system is great; but blind donkeys walking in circles to turn the upper on the nether millstone would be a lot less dull. In fact, on seeing that particular vision of the future—and make no mistake about it, those people mean that to *be* the future—Larry Niven had a suggestion. I should, he said, put on jack boots and revolver, and carry a whip; and we would find a gentleman of the black persuasion and dress him in rags and have him sit on the bicycle seat to grind our bread. It should, Larry mused, make a good photograph. A picture of the future.

I can't quarrel, except for details. The person seated on the bicycle seat might not be black, and might not be made; the person with whip might not be white or male; but if grinding one's corn to make one's bread requires that kind of labor, then slavery is not far away. In the sweat of thy face shalt thou eat bread; and mankind has been trying to get someone else to do the sweating ever since, and rather successfully at that. As a lark, as something chic, as a diver-

192

sion for the middle-class student spending a summer in an appropriate technology commune, labor-intensive systems are all very well; but as a necessity it gets *regular*; it is not amusing as a way of life.

Nor am I merely having a laugh at some silly people. There is a great deal to be said for conservation: but it is not a goal in itself. Look: why *shouldn't* we have heated swimming pools? What's *wrong* with big, comfortable, fast automobiles? Why is it evil to have throwaway flashlights, electric can-openers, warm houses in winter, air conditioning, patent medicines, luxury foods, electric typewriters, plastic models, fiberglass yachts with Dacron sails, pocket computers, my own postal scale in my office so I don't have to go down to the Post Office before mailing this manuscript —all the myriad conveniences, yea, luxuries of this marvelous modern civilization?

True, they may cost too much; we may not be able to afford wasteful items; and we may of necessity be forced to put away some of our luxuries. If so, then we must; but these new anti-technology intellectuals who have so much influence over the next generation would do it *gladly*. Look at Carter's energy policy. See Schlesinger on conservation. Look at the research budgets.

That brings us to my previous point. The trend is against technology and high energy; against development, and in favor of the "Small Is Beautiful," "Only One Earth" philosophy of the appropriate-technology movement. But surely, Pournelle, the Appropriate Technology movement is the best insurance against the knockout that so worries you? Making people self-sufficient, in small groups, building communes, conserving energy, taking care of one's own wastes, reducing the dependency on The System—

If you'll believe that you'll believe anything. Leaving out whether it's possible, either physically or politically, to insure against disaster by inducing large numbers of people to be "self-sufficient," if the Appropriate Technology movements succeed, *my* world will vanish because they *want* it to vanish. One of their goals is to suppress the kind of technology and development I want. They *like* "labor-intensive" industry.

The doomsters, neo-Malthusians, appropriate technology advocates, "ecologically concerned," and all the others, have set Zero-Growth as their goal; and my world is doomed if they succeed.

We need not envision either war or Vacca's knockout to imagine a world in which my vision of man's vast future remains the mere ravings of a science fiction writer. Merely continue as we are now: innovative technology discouraged by taxes, environmental impact statements, reports, lawsuits, commission hearings, delays, delays, delays; space research not carried out, never officially abandoned but delayed, stretched-out, budgets cut and work confined to studies without hardware; solving the energy crisis by conservation, with fusion research cut to the bone and beyond, continued at level-of-effort but never to a practical reactor; fission plants never officially banned, but no provision made for waste disposal or storage so that no new plants are built and the operating plants slowly are phased out; riots at nuclear plant construction sites; legal hearings, lawyers, lawyers, lawyers . . .

Can you not imagine the dream being lost? Can you not imagine the nation slowly learning to "do without," making "Smaller Is Better" the national slogan, fussing over insulating attics and devoting all attention to windmills; production falling, standards of living falling, until one day we discover the investments needed to go to space would be truly costly, would require cuts in essentials like food—

A world slowly settling into satisfaction with less, until there are no resources to invest in That Buck Rogers Stuff?

I can imagine that.

I even see trends in that direction. Mr. Carter has said no to plutonium, a decision we could live with; and followed that with an energy message that in a full hour had not one reference to the word "fusion," while out at Livermore and

Los Alamos they are laying off people whose entire professional lives have been spent in fusion research. Our President has told us we will have to make sacrifices, but he has given us nothing to sacrifice for. We shall insulate our attics, but mostly we shall use the energy crisis as a means for redistributing income and increasing taxes and increasing the bureaucracy. (And we shall penalize hell out of those who, like myself and Poul Anderson, long ago insulated and learned to keep our automobiles in tune ...)

Where is the innovation? The imagination? I expected a lot more from the President's energy message. I expected at the very least a massive research campaign: a Manhattan Project in agricultural research to develop plants capable of harnessing larger fractions of the solar energy falling on them; another to develop means for extracting energy and fertilizer from our sewage and trash; a specific plan to insure the safety of nuclear power plants while also assuring investors that the plants will be built, will not be unreasonably delayed by perpetual hearings and court challenges; perhaps a promise of restoration of some of the funds for fusion research; more funding for the ocean thermal energy system; something for the "slow" breeder, which uses the uranium-thorium cycle and doesn't produce any plutonium and can't be used to make bombs or terror weapons; something. Perhaps not all of the above, but something.

Instead we were promised an income-leveling tax system and told to tighten our belts while insulating our attics. Make do. Expect less. The "spree" is over. There's only one Earth ...

Now look: conservation is not going to get us to space. At best conservation can save us about half what is used for space heating: a few years' growth increase. There's nothing wrong with that, but there's nothing right about it either. I hate to say this, but the only problem with waste is that it's costly. Suppose, just suppose for a moment, that we suddenly discovered a million years' worth of fossil fuels. Better yet, suppose, just suppose, that we really had workable solar power systems of great efficiency such that they could supply us with all the power we ever wanted at trivial costs. Would it be worthwhile insulating the attic? Obviously not, unless it could be shown that an un-insulated attic was somehow harmful to the rest of us. There's nothing good *per se* about conservation, and nothing bad *per se* about throwaway cigarette lighters or Cadillacs. It happens that at the moment we may not be able to afford them and perhaps we'd best do without; but surely not-having-Cadillacs is not a positive goal? Surely not the only positive goal?

But aren't we going after solar power? And won't that ultimately solve all problems? Yes, to both; but we won't get it in time. Sorry: we may not get it in time. Solar power is risky and expensive technology. It is inevitable that some form of it will eventually power the Earth, but that may take far longer than Mr. Carter seems to believe.

Freeman Dyson: "In the very long run we must have energy that is clean and perpetual. We shall have solar power. In the long run we must have energy that is obtainable and available in large quantities. We shall have fusion. In the near term we must have energy that is now available. We have fission power. For the present we must have energy that is cheap, convenient, and easily obtained. We have coal, oil, and natural gas. Nature has been kinder to us than we had any right to expect."

I wish I were that confident; but I am not. The trends, in my judgment, do not augur well for us getting to the long run; and trying to skip the near term and long run and jump directly to the very long run is comparable, in my judgment, to Congress ordering Goddard to send a ship to the Moon by 1935 or give up those crazy rockets.

We could do it. We could spiral down until we have so few surplus resources that Roberto Vacca's knockout becomes possible; to a point where we have little, and many seethe with discontent, and suddenly it all explodes in riots, or war, or chaos; and when we recover from that (some of us) we will find that the business of living takes all our talents and energies; and our grandchildren will curse our memories.

* * *

It doesn't have to be that way. Here is another future.

First, war. Consider the following sequence of events. DeGaulle gives China the atom bomb, and when asked why says he has done nothing that Richelieu didn't do: when threatened with a European enemy, aid the Turks (or some other Asian). The Soviets begin a new Berlin crisis. The Chinese attack a small Soviet island at Ussuri. The Soviets back down on Berlin and begin moving troops in massive numbers toward their eastern frontier. After three weeks of buildup they retake the island. The Chinese glare at them across the river.

Marshal Gretchko goes to diplomatic parties and makes dark hints. In ten days there may be nuclear war. He hopes the West will understand. The West makes no response at all.

The Soviets discover their nuclear weapons are very dirty: following atomic war with China most of the population of Japan may die from the fallout. In the US knowledgeable people get out their Bendix fallout radiometers and dosimeters and buy new batteries for them and make a few other preparations.

More Soviet troops move to the east, a massive deployment until over a hundred divisions are on the Chinese border.

Henry Kissinger takes satellite photographs of the Soviet deployment to China. Ping-pong teams begin moving back and forth. Lin Piao, the most dangerous man in China, dies in mysterious circumstances that may never be known to Westerners.

Nixon goes to China.

It all happened, in that sequence. Add this: a few years

later the Soviets declassified their fusion research and brought the bag over here in the hopes that we could make use of it. (We immediately classified what they gave us, putting a blanket over the blackboard.) Soviet experts privately say their need for fusion energy is great; they have a lot of development to do. They also build a fast breeder fission reactor based on US technology (we have yet to build one, of course).

Suggestive of what? This much: that at least some officials in the Soviet Union are now convinced that games against nature have a higher payoff than conquest; that they couldn't run China when they had it, and can't now; that if their system were introduced into western Europe (Hungarian joke: "The Soviets have crossed a cow with a giraffe to produce a marvelous animal that feeds in Budapest and is milked in Moscow") European production would not only fall, but the *Soviet* economy would be worse off.

Give it a couple of generations and possibly, just possibly, the dinosaurs will die. It takes skillful diplomacy by the West, but it's possible. If we can discourage the dinosaurs until the technicians are in control of the Soviet Union, we will have peace. My private opinion is that the best way to discourage the dinosaurs is to remain so strong that they have no expectation of winning—Cato's advice. "If you would have peace, prepare thou then for war." I realize that is not universally accepted, neither among the ruling elite in the US, nor among you, my readers.

But assume the dinosaurs are contained, and the trends toward cooperation continue; that we do not end in war. It is possible.

Next, the energy crisis. We know how to produce the energy. Solar Power Satellites could be delivering power to Earth well before the end of the century. Geopressure zones can relieve the natural gas crisis. At the AAAS meeting in 1978 the fusion scientists announced that we could, given money and luck, have a working reactor delivering commercial power before 1995; it isn't inevitable, but it's certainly possible.

Princeton University recently announced that through the use of highly advanced technology—neutral beam currents—they were able to achieve temperatures of 60 million degrees. This unexpected breakthrough (it wasn't thought they'd do that well for several more years) could lead to a working fusion reactor before 1990—if we want one.

And after all—as the dollar plummets, it is going to become obvious to all that when your energy system is to pay $50 billion and more each year for foreign oil, you can afford *any* alternative; that research is cheap at the price. Perhaps the western states' senators will exert enough pressure to get the fusion budget restored. After all, Carter backed down on canceling all those waterways plans; and it's even possible that someone will appreciate the need for research without regard to electoral politics. Isn't it?

We could have working reactors before the year 2000.

Then there are the ocean thermal systems: at the moment they're getting paper-study money only, and although it's widely announced that a demonstration plant will be built, the fact is that nothing beyond paper has been funded; but suppose the money comes through to bend tin and cut metal, and the plant is built. It won't be commercial, but from it we should learn how to make commercially viable plants; and we can build research stations as described in my book HIGH JUSTICE.

And—we could go to space. This year the Congress asked, pointedly, why NASA proposed no new starts; why there was nothing bold and imaginative in the national space program. The administration may have bought Zero-Growth, and true, Vice President Mondale while a Senator each year introduced a bill to kill NASA entirely, abolish all its research and development; but the Congress seems to have taken new interest in space. Only one of the shuttles has been canceled. We do have a shuttle commitment, if only because we have treaty obligations to the Europeans who have developed Spacelab. The *Enterprise* will never go to space, but others will. It is possible that commercial firms will be given the opportunity to rent time in orbit at reasonable rates.

It has been the historic role of government to explore the frontiers and build roads to them; perhaps the Congress will recognize that space is not essentially different from California in that respect.

There is so much going for us. Biologists are fairly sure that a massive research effort will double the efficiency of plants. The computer engineers continue to produce their marvels. We have discovered geopressure zones.

And we can afford the research: indeed, given the alternative of $50 billion annually now, and $100 billion a year expected, as the price of imported fuels, we can afford *everything:* we can pursue all our lines of research, fund them well, and confidently expect to save vast sums when even *one* of the many routes to energy independence pays off.

There is an alternative to Zero-Growth. People could learn to expect *more,* not less; know that if they work harder they can have it, for themselves and their children.

It's a possible future. Isn't it? If it's not, don't tell me; because I want to believe that there is a chance; that I may live to see that world I describe in my lectures; that before I die I can say "My generation gave mankind the planets and the stars; and I was a part of it."